Cladistics

Systematics Association Publications

1. Bibliography of key works for the identification of the British fauna and flora, *3rd edition* (1967)†
   Edited by G. J. Kerrich, R. D. Meikle, and N. Tebble
2. Function and taxonomic importance (1959)†
   Edited by A. J. Cain
3. The species concept in palaeontology (1956)†
   Edited by P. C. Sylvester-Bradley
4. Taxonomy and geography (1962)†
   Edited by D. Nichols
5. Speciation in the sea (1963)†
   Edited by J. P. Harding and N. Tebble
6. Phenetic and phylogenetic classification (1964)†
   Edited by V. H. Heywood and J. McNeill
7. Aspects of Tethyan biogeography (1967)†
   Edited by C. G. Adams and D. V. Ager
8. The soil ecosystem (1969)†
   Edited by H. Sheals
9. Organisms and continents through time (1973)†
   Edited by N. F. Hughes
10. Cladistics: a practical course in systematics
    P. L. Forey, C. J. Humphries, I. J. Kitching, R. W. Scotland, D. J. Siebert, and D. M. Williams

† Published by the Association (out of print)

Cladistics

A Practical Course in Systematics

Peter L. Forey, Christopher J. Humphries,
Ian J. Kitching, Robert W. Scotland,
Darrell J. Siebert, and David M. Williams.

CLARENDON PRESS · OXFORD

Oxford University Press, Walton Street, Oxford OX2 6DP
Oxford New York Toronto
Delhi Bombay Calcutta Madras Karachi
Kuala Lumpur Singapore Hong Kong Tokyo
Nairobi Dar es Salaam Cape Town
Melbourne Auckland Madrid
and associated companies in
Berlin Ibadan

Oxford is a trade mark of Oxford University Press

Published in the United States
by Oxford University Press Inc., New York

First published 1992
First published in paperback 1993
Reprinted 1994 (twice)

A catalogue record for this book is available from the British Library

Library of Congress Cataloging in Publication Data
(Data available)

ISBN 0 19 857766 4 (Pbk)

Printed and bound in Great Britain by
Biddles Ltd, Guildford & King's Lynn

Preface

Cladistics: a practical course in systematics originated as course material for a workshop sponsored by the Systematics Association in 1991. Stephen Blackmore, Keeper of Botany, and Laurence Mound, Keeper of Entomology, of the Natural History Museum in London realized that the course manual would be in demand and encouraged its publication. The Systematics Association then agreed to issue the manual as part of their Systematics Association Publications series.

Considerable interest was shown in the workshop *Cladistics: Theory and Practice* which was held in 1991 at the Natural History Museum, and it was oversubscribed to such an extent that the Systematics Association decided to sponsor it again in 1992, and perhaps on a continuing basis after that. The present version of the workshop manual incorporates new material and revisions undertaken in preparation for the 1992 workshop.

The intent of the workshop was to cover modern cladistics in an intense, one-week short-course, and to have it taught by expert, practising cladists. For the 1992 workshop Robert Scotland prepared theoretical background and character-coding materials; Ian Kitching prepared materials on character polarity and algorithms; Darrell Siebert prepared materials on tree statistics and alternatives to parsimony; David Williams prepared materials on molecular approaches and prepared all of the figures; Peter Forey prepared materials on fossils and classification; and Chris Humphries prepared materials on biogeography. Chris Humphries, Darrell Siebert, and David Williams edited the manual. Darrell Siebert prepared the index.

Many individuals and institutions contributed directly, or indirectly, to this manual. Paddy Orchard arranged a generous grant from the training budget of the Natural History Museum; Richard Cloutier presented materials on fossils during the first workshop; John Benfield gave substantially of his time behind the scenes; and the students of the first workshop made many helpful observations and comments. Most importantly, the Systematics Association Council provided financial support and guidance throughout.

C.J.H.
D.M.W.
D.J.S.

London
January 1992

Contents

Authors

Peter L. Forey
Department of Palaeontology, The Natural History Museum, London

Christopher J. Humphries
Department of Botany, The Natural History Museum, London

Ian J. Kitching
Department of Entomology, The Natural History Museum, London

Robert W. Scotland
Department of Plant Sciences, University of Oxford, Oxford

Darrell J. Siebert
Department of Zoology, The Natural History Museum, London

David M. Williams
Department of Botany, The Natural History Museum, London

Introduction

Darrell J. Siebert

Systematics underpins all of biology. It is the discipline through which comparative biology progresses, whether the subdiscipline of interest is ecology, biogeography, evolution, or physiology. Yet the teaching of systematics is in decline in the UK. The situation has reached such a state that the House of Lords Select Committee on Science and Technology (Sub-committee II — Systematic Biology Research) has taken evidence, the Linnean Society of London and the Systematics Association jointly have sponsored a meeting, the Natural Environment Research Council has formed a review committee, and the British Ecological Society held a meeting, all about the future of systematics. Almost everyone involved agrees that something has gone wrong. At a time of accelerating environmental crises, when the need for the expertise of taxonomists has never been greater, taxonomists seem to be in short supply, and few institutions are training them.

The Systematics Association, pursuant to its mission of teaching systematics and taxonomy, decided to sponsor a short, intense workshop on cladistics. The aim was to cover modern cladistics, and to have it taught by expert, practising cladists. The tertiary education level was chosen as the level at which the course should be taught and it was decided to make it as practical and 'hands-on' as possible. Thus, the course consists of about ten hours of lectures, for which this manual is background material, and about 30 hours of practicals devoted to familiarization, by way of data analyses, with computer programs such as COMPONENT, Hennig86, PAUP, and parts of PHYLIP.

The attempt to redress a decline in the teaching of systematics and taxonomy with a course on cladistics might seem controversial, especially with its emphasis on parsimony. However, cladistics has become the preferred tool for comparative biology. Systematists the world over are turning to it as the method of choice for comparative studies (Hull 1988). Cladistics is not necessarily the hyper-modern development it sometimes has been made out to be. Some (Nelson and Platnick 1981) have suggested cladistics has always been part of systematics. Modern cladistics, however, effectively dates from Hennig (1966). This recent codification has allowed important insights into the workings of systematics. Perhaps the most important of these are systematic relationships, based on homology, and natural, monophyletic groups. Groups recognized by our predecessors that we still recognize as valid today are those that are monophyletic and are characterized by homologies. Assemblages that have fallen by the wayside or are continual sources of discussion, for example reptiles, cannot be characterized by homologies and are not monophyletic.

Although simple in theory, cladistics is sometimes difficult to practise, though no more so than any other sort of systematics. The chapters that follow present the practice of cladistics in outline. A general background as an introduction to parsimony in systematics is presented first. Then character coding and the topic of polarity is discussed. Parsimony methods follow, then a discussion of algorithms and the search for the most parsimonious branching diagram. Tree statistics, alternatives to parsimony in systematics, consensus trees, character conflict, molecular applications, and the relevance of fossils come next. By this point the basics of cladistics have been covered. Two special topics follow: cladistic biogeography, and the implementation of cladistic results in classifications. Both are designed to show how cladistics is put to use; cladistic systematics leads to a general evolutionary theory of Life and Earth history and is applied to everyday taxonomy through classification. Finally, an extensive reference list is provided. Any reader can follow any of the above topics to its present limits by way of these references. The limits of some of these topics, such as biogeography, are moving fast. A reader will have to be aware of further developments to remain current. This is no small chore, as the development of cladistics is occurring across all subdisciplines of biological systematics. A few journals, however, such as *Cladistics*, *Journal of Biogeography*, *Systematic Biology*, and *Taxon*, do specialize in the topics of systematics, thereby making the job of keeping up-to-date easier.

Taken together, these topics are intended to make cladistics more widely available and understood by a greater proportion of the biological community. Cladistics is a powerful systematic discovery procedure, and its use is rapidly becoming more widespread. An important factor in its rapid spread is its codification. The procedures of cladistics, in explicit form, are available to all, and computer programs to implement them are inexpensive. This means that comprehension of cladistic classification is open to all and that many more people are potentially able to contribute to the formulation of its methods.

1.
Cladistic theory

Robert W. Scotland

1.1 INTRODUCTION

In biology, cladistics is a method of systematics (Patterson 1980), most co-herently formulated by Hennig (1950, 1966), which is used to reconstruct genealogies of organisms and to construct classifications. However, it is also a general approach to classification which can be used for organizing any com-parative information, having been independently discovered in linguistics (Platnick and Cameron 1977; Bonheim 1990) as well as being used in bio-geography (see Chapter 9). Explanation, discussion and exploration of the cladistic method and theory can be found in Patterson (1978, 1980, 1981*a*, 1982*a*), Eldredge and Cracraft (1980), Nelson and Platnick (1981), Wiley (1981), Janvier (1984), and many journals, especially *Cladistics* and *Systematic Zoology* (now *Systematic Biology*), while an appreciation of cladistics within the wider context of the history of systematics, can be found in Nelson and Platnick (1981). The axioms of cladistics are:

1. Nature's hierarchy is discoverable and effectively represented by a branching diagram.

2. Characters change their status at different hierarchical levels. Characters within a study group that are either present in all members of the study group or have a wider distribution than the study group (plesiomorphies) cannot indicate relationships within the study group.

3. Character congruence is the decisive criterion for distinguishing homology (synapomorphy) from non-homology (homoplasy).

4. The principle of parsimony maximizes character congruence.

1.2 FORM: HOMOLOGY AND ANALOGY

By way of introduction, consider Richard Owen's distinction between homology and analogy. Owen (1843) attempted to distinguish 'true' similarity (homology) from superficial similarity (analogy). Homology was considered to be the signal for discovering the natural hierarchy, whereas analogy was considered to be noise and misleading. Owen defined homology as 'the same organ . . . under every variety of form and function' and analogy as 'a part or organ . . . which has the same function' but derived from a non-homologous base.

Established analogues are the wings of a bird and of a butterfly. In contrast, the front leg of a horse, the human arm and the wing of a bird are viewed as homologues as they possess similar but different structures in the same topographical positions. In this context, it is useful to consider whether the wings of birds and bats are homologous or analogous. The most commonly accepted classification of tetrapods accepts that birds (Aves) are more closely related to crocodiles (Crocodylia) than to bats (Mammalia) (Nelson and Platnick 1981). At this hierarchical level, the 'wingness' of the forelimb in birds and bats is analogous, but at a hierarchical level including all tetrapods, the wings of birds and bats are viewed as homologous, but as forelimbs, not as wings. Their superficial similarity as wings does not stand up to rigorous investigation. For example, the digits of the wings and their coverings are different, either being feathers (birds) or skin (bats). This example demonstrates that statements of putative homology demand three conditions to be met: (1) there must be a specifying phrase, homologous to what?; (2) the hierarchical level at which a homology is being postulated needs to be stated; (3) the proposed similarity must stand the test of rigorous anatomical comparison.

1.3 SPECIAL SIMILARITY

Hennig (1966) recognized that similarities form an important part of any taxonomic study. However, he reasoned that the features (characters that form the basis of any classification) change their status relative to the taxonomic problem being investigated, and that similarities are therefore of three sorts, two of which concern us for the moment and the third is described in section 1.4 below. Hennig differentiated primitive similarity from advanced similarity, and called these plesiomorphy and apomorphy respectively. His starting point was that evolution had occurred and that shared characters between organisms were the result of shared inheritance. He further reasoned that only shared derived characters (synapomorphies) are evidence of exclusive common ancestry (strict monophyly, see below).

The example in Fig. 1.1 illustrates the concepts of plesiomorphy and apomorphy relative to four taxa. Taxa 2, 3, and 4 are taken as the study group with taxon 1 having the primitive character states (a, b, c, d), representing the most recent common ancestor of 2 (A, B, c, d), 3 (A, b, C, D) and 4 (A, b, C, d) with upper case letters representing derived character states. At this stage, it is not important how primitive and advanced characters are determined (see Chapter 3). Figure 1.1 illustrates that the advanced character states (A) and (C) diagnose groups comprising (2 3 4) and (3 4) respectively. The shared presence of character state (b) in taxa 3 and 4, within the study group, cannot confirm

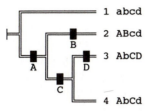

Fig. 1.1 Plesiomorphy and apomorphy: taxa 2–4 comprise the ingroup; plesiomorphic character states are represented in lower case letters, apomorphic character states in upper case letters. Apomorphic character state A diagnoses taxon (2 3 4), C diagnoses taxon (3 4). Character states b and d are plesiomorphic; b cannot confirm and d cannot contradict taxon (3 4).

the group comprising (3 4), as character state (b) is a retained primitive character (plesiomorphy) which is present also in taxon 1. The shared presence of character state (d) in taxa 2 and 4, within the study group, does not contradict the group comprising (3 4) because it is a retained primitive character. An example from animals of (b) and (d) type plesiomorphic characters is the presence of five toes in humans, hedgehogs, lizards and frogs. The presence of five toes on the back legs in these animals is evidence that they are related by common ancestry as they belong to a group comprising four-footed, five-toed vertebrates (tetrapods). For at that level (tetrapods) a five-toed foot is an advanced feature, when compared with the fins of fishes. However the rear legs of some tetrapods do not have five toes, for example, horses (one), cattle (two) and birds (four) because they have undergone modifications of the five-toed condition. Therefore, given the problem of classifying humans, frogs and horses, the shared presence of five toes between humans and frogs is irrelevant as it is a retained primitive character (plesiomorphy). Therefore, the status of a character changes depending on the hierarchical level being considered, such that a plesiomorphic character at a particular hierarchical level is a synapomorphy of a less restricted level.

1.4 CONFLICTING SIMILARITIES AND PARSIMONY

The third type of similarity is termed homoplasy. The hypothetical example shown in Fig. 1.1 is unproblematic because there are no conflicting characters. This is seldom the case in the real world. If the earlier example is expanded to include two further characters each with two character states (e, E) and (f, F), with lower case letters representing primitive character states and upper case derived, conflict then results. The distribution of all six characters is shown in Table 1.1. Excluding taxon 1, as it is the immediate common ancestor of 2, 3, and 4, there are three possible resolved systematic arrangements of taxa 2, 3, and 4. These three solutions are diagrammatically illustrated in Fig. 1.2.

Given the three possibilities (Fig. 1.2) what criterion can be employed to choose between them? To explore this further, consider each of the three possibilities. Figure 1.3a illustrates the solution that taxa 3 and 4 are more closely related to each other than either is to taxon 2, represented in parenthetical notation as (1(2(3 4))). The agreement between the solution and the characters (data), can then be assessed by mapping the distribution of characters on to the cladograms, Fig. 1.3a–c. Similarly, the other two possible solutions (1(3(2 4))) and (1(4(2 3))) are illustrated in Fig. 1.3b and 1.3c respectively.

Table 1.1 Distribution of six characters (A–F) for four taxa (1–4); presence of a character in a taxon is indicated by (+). Data for Fig. 1.3.

Taxa	Characters					
	A	B	C	D	E	F
1	–	–	–	–	–	–
2	+	+	–	–	–	+
3	+	–	+	+	+	+
4	+	–	+	–	+	–

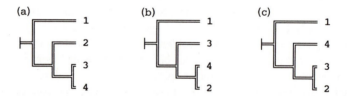

Fig. 1.2 Diagram of the three possible fully resolved solutions for the ingroup taxon (2 3 4).

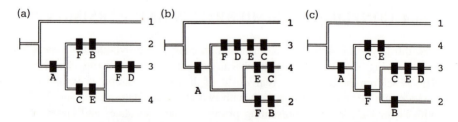

Fig. 1.3 Diagram of the three possible fully resolved solutions for taxa 2, 3 and 4 with characters A to F (Table 1.1) mapped onto the branches. (a) Tree with taxa 3 and 4 as closest relatives. (b) Tree with taxa 2 and 4 as closest relatives. (c) Tree with taxa 2 and 3 are closest relatives.

How then can a choice be made among these three solutions? A criterion to solve this type of problem is the concept of the simple hypothesis (often attributed to William Ockham, ca. 1280–1349), the criterion of parsimony for choosing between competing scientific hypotheses (Camin and Sokal 1965). The parsimony principle reasons that, given several solutions to a problem, the most economical should be accepted, or in other words, the most parsimonious solution is to be preferred to all others. This approach allows the choice of Fig. 1.3a, as it is seven steps long (a step is equivalent to an occurrence on the diagram), whereas Fig. 1.3b and 1.3c are nine and eight steps long respectively. The parsimony criterion serves to maximize character congruence (see below).

1.5 THE RELATIONS BETWEEN SIMILARITIES

There are five ways in which two characters may relate to one another concerning their ability to specify groups. Consider character A that is shared by three organisms 1, 2, and 3 (1 2 3) (Fig. 1.4a). A second character, B, is present in organisms 4 and 5, in which case it specifies a different group (4 5) (Fig. 1.4b). Character C is present in organisms 2 and 3, which is a sub-set (2 3) of the initial group (1 2 3) (Fig. 1.4c). Character D is found in organisms 1, 2, and 3, and therefore specifies the same group as character A (1 2 3) (Fig. 1.4d). Lastly, a character, E, is present in organisms 3 and 4 and it therefore

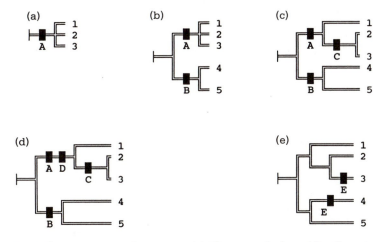

Fig. 1.4 Relations among characters. (a) Character A shared by three taxa, 1, 2 and 3. (b) Character B is present in taxa 4 and 5, specifying a different group. (c) Character C is present in taxa 2 and 3, which is a sub-set of the initial group. (d) Character D is present in taxa 1, 2 and 3, specifying the same group as character A. (e) Character E, present in taxa 3 and 4, specifies a group (3 4) which conflicts with the initial groups (1 2 3) and (4 5). Relative to A, D is congruent, B and C are consistent and E is incongruent.

specifies a conflicting group (3 4) to the initial groups (1 2 3) and (4 5) (Fig. 1.4e). Therefore, relative to an initial group (1 2 3) based upon character A, character B specifies a different group, character C a sub-set of the initial group, character D the same group, or character E a conflicting group. Relative to character A, characters B and C are consistent, D is congruent, and character E is in conflict.

Figure 1.5a,b contrasts two possible branching diagrams for five taxa. Figure 1.5a is six steps long with one incongruent character (E) and is the most parsimonious solution. Figure 1.5b is nine steps long and has four incongruent characters (A, B, C, D). The acceptance of the parsimony criterion for the evaluation of competing systematic hypotheses, therefore, serves to maximize the congruence of characters and to minimize character conflict when incongruency is present.

1.6 HOMOPLASY AND THE INTERPRETATION OF CHARACTER CONFLICT

Homoplasy can thus be of two sorts. First, as in the bird and bat example, what are first thought to be similarities, on close comparison, turn out not to be so. This has been termed analogy, or in evolutionary terms, convergence. In other cases, it is possible for anatomical singulars (Riedl 1979), ('the same organ . . .', Owen 1843) to be incongruent and this can be termed parallelism. Patterson (1982a, 1988a) reviewed the literature and discussed criteria to distinguish different types of homology from homoplasy. The view adopted here is that homoplasy is incongruent data that cannot be further subdivided (Platnick 1979).

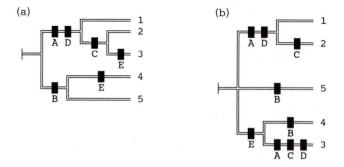

Fig. 1.5 Topology, parsimony and congruence. (a) Cladogram of 5 taxa and 5 characters, of which character E is incongruent; the cladogram is 6 steps long. (b) An alternative, less parsimonious topology for the same taxa and characters. This topology requires 9 steps; characters A, B, C and D are incongruent.

An example of character conflict leading to different, but equally parsimonious, cladograms is given in Fig. 1.6a,b. The distribution of character A for three taxa 1, 2, and 3 is illustrated in Fig. 1.6a. This distribution can be explained as two separate, parallel steps on the diagram. However an equally parsimonious explanation of this character is shown in Fig. 1.6b, where the character is placed as diagnostic of three taxa and is interpreted as secondarily lost in taxon 3. Therefore both explanations are equally parsimonious but involve different interpretations of character change. In other words, the difference between parallelism and reversal may be merely one of interpretation.

1.7 MONOPHYLETIC, PARAPHYLETIC, AND POLYPHYLETIC GROUPS

Patterson (1982*a*) synonymized homology with synapomorphy and in so doing coupled the major concept of comparative anatomy, namely homology, with the historical perspective of Hennig's (1966) special similarity (Farris 1977*a*). The justification in systematics of the common plan, which had shifted from Owen's archetype to Darwin's common ancestor more than 100 years ago, finally came into sharper focus because Hennig argued that common ancestry alone is no justification for establishment of systematic groups.

Hennig (1966) recognized that all organisms may reasonably be assumed to be related through common ancestry. He therefore argued that common ancestry, in itself, was not a rigorous criterion for the justification of taxonomic groups. In other words, if life arose once and is monophyletic (see below), then all of life is related through common ancestry. Within the context of evolutionary taxonomy, prior to Hennig (1966), a monophyletic or natural group included taxa that shared common ancestry, the consequences being that any group was monophyletic. Hennig (1966) defined monophyly more precisely and more restrictively than before, such that taxon A is more closely related to taxon B than C or D, if A and B have a common ancestor that is not also an ancestor of C or D. With this definition of strict monophyly, Hennig (1966) distinguished

Fig. 1.6 Interpretation of conflicting, equally parsimonious character distributions. (a) Distribution of character A for taxa 1, 2 and 3. This distribution is explained by two separate, parallel steps on the diagram. (b) An equally parsimonious explanation of character A; chracter A is diagnostic for (1 2 3) and is interpreted as secondarily lost in taxon 3.

between three types of groups: monophyletic, paraphyletic, and polyphyletic (Fig. 1.7a–c). Along with Hennig's (1966) increased precision with regard to monophyly, the concept of paraphyletic groups was new.

For some (Patterson 1981*a*; Nelson 1989), the recognition and elimination of paraphyly was Hennig's most significant and lasting contribution to systematics. Hennig (1966) defined paraphyletic groups as being diagnosed by plesio-morphies and not including all descendants of a common ancestor. For example, fishes (Pisces), recognized by the presence of paired fins, are paraphyletic because the paired limbs of tetrapods are most parsimoniously interpreted as modified paired fins, and because some fishes share with tetrapods synapo-morphies that are not present in other fishes (jaw bones of bony fishes, choanae of lungfishes). Other examples of paraphyletic groups, which have so far resisted diagnosis by any synapomorphies, are invertebrates, reptiles, algae, and gymnosperms.

A polyphyletic group was defined by Farris (quoted in Wiley 1981, p. 84) as: 'a group in which the most recent common ancestor is assigned to some other group and not to the group itself'. There is general agreement that polyphyletic groups which are based on convergent characters, for example wings (birds and bats) have no place in systematics.

For Hennig (1966), monophyletic, paraphyletic, and polyphyletic groups are defined in terms of their relation to common ancestry. However, for Platnick (1979) and Patterson (1982*a*) this involved faulty logic as the criterion for group membership relied upon an appeal to something that is unknown, namely common ancestry. The problem was that theories about characters (synapo-morphies) and theories about groups (monophyly) both appealed to common ancestry for justification. Platnick (1979) and Patterson (1982*a*) reasoned that theories of characters (homologies) provide evidence of groups which can then be interpreted in terms of common ancestry. Thus the role of ancestry was removed from analysis and became part of the narrative to explain patterns.

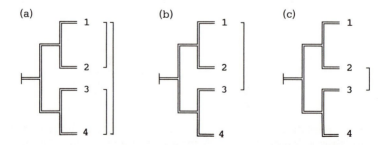

Fig. 1.7 Diagram of the three types of systematic groups. (a) Groups (1 2), (3 4), and (1 2 3 4) are all monophyletic. (b) Group (1 2 3) is paraphyletic. (c) Group (2 3) is polyphyletic.

Therefore, as represented above (Fig. 1.7), monophyletic groups are discovered and paraphyletic and polyphyletic groups resolved in terms of character distribution and hierarchy, but free from *a priori* notions of common ancestry.

1.8 SISTER GROUPS AND ANCESTOR–DESCENDENT RELATIONSHIPS

Hennig (1966) defined sister groups as: 'species groups that arose from the stem species of a monophyletic group by one and the same splitting process' and further stated: 'once a monophyletic group has been recognised, the next task of phylogenetic systematics is always to search for the sister group'. Figure 1.8 is a branching diagram indicating the relationships among five taxa. Taxa 1 and 2 are sister groups, the monophyletic group (1 2) is sister group to taxon 3, the monophyletic group (3(1,2)) is sister group to taxon 4 and the monophyletic group (4(3(1,2))) is sister group to taxon 5.

Phylogenetic relations between taxa can be seen as sister groups, i.e. taxa sharing a common ancestor (Fig. 1.9*a*), or as ancestor-descendent relations (Fig. 1.9*b*), in which taxon B is the direct ancestor of taxon A. The problem with this latter aspect of phylogenetic relationship is that there is no empirical method for detecting ancestors from their descendants because there is only one criterion for group recognition, homology (synapomorphy), and this cannot distinguish ancestor–descendent relationships.

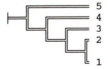

Fig. 1.8 Sister group relationships: taxon 3 is the sister group of taxa 1 and 2, ((1 2) 3); taxon 4 is the sister group of taxa 1, 2 and 3 (((1 2) 3) 4); and taxon 5 is the sister group of taxa 1, 2, 3, and 4 ((((1 2) 3) 4) 5).

Fig. 1.9 Phylogenetic relations. (a) Taxa A and B seen as sharing a common ancestor. (b) Taxa A and B seen in an ancestor-descendent relationship; taxon A is the direct ancestor of taxon B, or *vice versa*.

1.9 THE TRANSFORMATION OF CLADISTICS

Originally, Hennig (1966) formulated the cladistic method relative to an evolutionary model in which speciation was dichotomous and species went extinct at nodes. Various authors reasoned that such a model was not a necessary part of cladistic method. For some, Nelson (1978, 1979, 1985, 1989), Nelson and Platnick (1981), Brady (1982), Patterson (1981*a*, 1982*a*, 1983, 1988*a*), Rosen (1982, 1984), Janvier (1984), and Rieppel (1985, 1988), cladistics is a discovery procedure independent from *a priori* considerations of phylogeny. Patterson (1988*b*) went further and stated: 'If the causal explanation of pattern is to be convincing and efficient, the pattern is better not perceived in terms of the explanatory process. This is the basic argument behind pattern cladistics'.

1.10 PHENETICS

A complete critique of phenetics (numerical taxonomy) is not attempted here, although it is pertinent that three criticisms raised of phenetics are mentioned briefly. Phenetic branching diagrams (phenograms) are constructed from a data matrix via a similarity matrix and a clustering algorithm. Similarity matrices can be of many different kinds, as can the clustering algorithms for creating the phenogram from the similarity matrix (Sneath and Sokal 1973). Many different topologies of phenogram are possible for the same data matrix depending upon the choices made in the similarity matrices and clustering algorithms. Therefore, one of the initial aims of phenetics, which was to create objective classifications and ultimate stability, has not been fulfilled as the results depend on the choice made from a wide variety of algorithms. Phenograms are made up of terminal taxa called operational taxonomic units (OTUs) which cluster together at different hierarchical levels determined by overall percentage similarity. As such, the relationships between individual characters and a phenogram cannot be interpreted (see Chapter 7). Many phenetic clustering algorithms group (cluster) jointly on absences and presences. This practice results in the recognition of groups (clusters) which include shared absence characters that are present in other OTUs. In other words, a taxon that has accumulated many autapomorphies clusters further away from other taxa because of the autapomorphies. This has the parallel effect of grouping (clustering) together taxa in which these autapomorphies are absent (see also Chapters 5 and 7).

1.11 ECLECTIC OR EVOLUTIONARY TAXONOMY

Detailed discussions of the many different positions held by evolutionary taxonomists can be found in Simpson (1961), Mayr (1969), Cronquist (1988) and Bock (1989). Evolutionary taxonomists often recognize paraphyletic (grade) groups as 'natural'. This is argued from the position of having to balance similarity with genealogy in the construction of a classification. This position holds that as evolutionary rates between taxa can be highly variable, taxa that have 'evolved further', as evidenced by a large number of autapomorphies, warrant special recognition. This is illustrated in Fig. 1.10a,b, which contrasts a cladistic with an evolutionary interpretation of the same data set. Another related criticism often levelled at evolutionary taxonomy is that it uses a narrative approach for constructing classifications, in which character distribution and data are not explored for equally parsimonious solutions and the classification results solely from the expertise and authority of the individual worker.

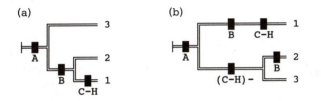

Fig. 1.10 Contrasting cladistic and evolutionary taxonomy classifications of the same data set. (a) The most parsimonious branching diagram for the given data. (b) A less parsimonious solution favouring an alternative arrangement of the terminal taxa and 'forcing' support for a paraphyletic group; taxa 2 and 3 are depicted as most closely related because they are similar in lacking characters C-H.

2.
Character coding

Robert W. Scotland

2.1 CHARACTER TYPES

Morphological characters can be of two kinds, either discrete (qualitative) or continuous (quantitative). This chapter deals exclusively with the coding of discrete data for qualitative characters. For example, the presence of either two or four stamens in a group of plants is a qualitative character with two discrete states. By comparison, the length of corolla tubes for the same group of plants may vary between 1 cm and 10 cm, with the lengths of individual corollas overlapping within and between individuals of a population and between species. These are continuous variables.

The problem with all characters, but particularly with continuous variables, is the question of whether they are cladistically significant or not, and how they might be coded into discrete states. The latter subject is not considered here, but see Pimentel and Riggins (1987), Cranston and Humphries (1988), Chappill (1989), Felsenstein (1988a), and Stevens (1991) for detailed discussions.

2.2 BINARY CHARACTERS

Consider the simple problem of how to code two plant species in which species A has four stamens and species B has two stamens. This information can be coded into simple binary form (Table 2.1), in which the condition in A is represented by 0 and that of B by 1. It is important to note that there is no particular meaning regarding the assignment of 0 or 1 to a particular stamen number, as the coding could easily be reversed, i.e. 0 = 2 stamens and 1 = 4 stamens, and retain the same meaning.

Table 2.1 Binary coding for two taxa for the character stamen number (see text)

Taxa	Character	Code
A	4 stamens	0
B	2 stamens	1

2.3 MULTISTATE CHARACTERS

In a similar example, the number of stamens for five taxa A, B, C, D, and E is 1, 2, 3, 4, and 5, respectively, and this can be coded as a multistate character as shown in Table 2.2. Multistate characters have more than two character states and are coded by integers equalling the number of character states.

2.4 TRANSFORMATION BETWEEN CHARACTER STATES

Three plant species A, B, and C with 2, 3, and 4 stamens, respectively, can be coded as a multistate character (Table 2.3). Characters coded in this way imply that a character can undergo transformation, i.e. the character of stamen number has undergone modification from one state to another. For the example given in Table 2.3, which has three character states, there are nine possible ways in which the transformation of three character states may be related (Fig. 2.1a–c). In an analysis incuding a multistate character with three states it is conceivable to permit all nine transformations, as shown in Fig. 2.1a–c, or limit these nine to fewer options through choices about character order and character polarity, as discussed below.

Table 2.2 Additive coding for five taxa for the multistate character stamen number

Taxa	Character	Code
A	1 stamen	0
B	2 stamens	1
C	3 stamens	2
D	4 stamens	3
E	5 stamens	4

Table 2.3 Additive coding for three taxa for the character stamen number

Taxa	Character	Code
A	2 stamens	0
B	3 stamens	1
C	4 stamens	2

2.5 UNORDERED AND ORDERED CHARACTERS

Consider that the relationships between three character states are known, in the sense that the intermediate step of the transformation can be determined. For example, the gain or loss of stamen number may be viewed as incremental and the gain or loss of two stamens proceeds via the intermediate step of gaining or losing one stamen. In such a case this additional information can be included in the analysis by ordering the character.

If the character is treated as unordered, then any state can transform into any other state with equal cost and any of the nine transformations shown in Fig. 2.1a–c are equally possible.

If treated as an ordered character a choice is made limiting the nine possibilities to a set of three depending upon the determined order. The three possibilities if the order is for incremental stamen gain or loss are shown in Fig. 2.2. The three possible transformations shown in Fig. 2.2 can be represented in one form as shown in Fig. 2.3.

Ordering a multistate character a priori determines the relationship of each character state with every other character state but is silent about the direction of transformation.

2.6 ADDITIVE BINARY CODING

In addition to coding different states within a column, multi-column coding can be utilized for character analysis. Additive binary coding is another way of coding ordered multistate characters, although the actual coding is more tedious and the output can be difficult to interpret (Swofford 1990). The character ordered in Table 2.3 as 0-1-2 can be recoded in additive binary form as 00-01-11 (Table 2.4).

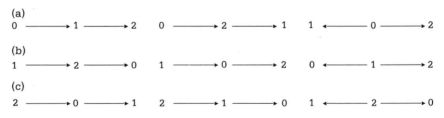

Fig. 2.1 The nine possible transformations among three character states.

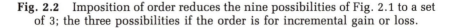

Fig. 2.2 Imposition of order reduces the nine possibilities of Fig. 2.1 to a set of 3; the three possibilities if the order is for incremental gain or loss.

2.7 BRANCHED CHARACTER STATE TREES

Although only linear transformed characters have been considered thus far, it is possible to include ordered branched characters in an analysis. If the transformation between four character states was determined a priori to be that shown in Fig. 2.4, this information can be included in an analysis. PAUP v.3.0 allows direct input of 'character state trees' within the user-defined character types option. To be used in Hennig86 the branched character (Fig. 2.4) would have to be recoded in additive binary form (Table 2.5).

0 ——————— 1 ——————— 2

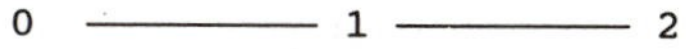

Fig. 2.3 The three possible transformations from Fig. 2.2 represented in one form.

Table 2.4 Coding multistate characters. For stamen number code 1 is additive, code 2 is additive binary coding

Taxa	Character	Code 1	Code 2
A	2 stamens	0	00
B	3 stamens	1	10
C	4 stamens	2	11

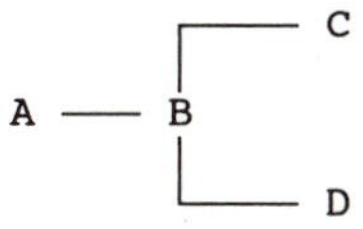

Fig. 2.4 Transformation between four character states determined prior to an analysis; the branched character can be coded in additive binary form (Table 2.5).

Table 2.5 Additive binary coding for the branched character state tree of Fig. 2.4. Taxa are identified as individual character states (A–D)

Taxa	Code
A	000
B	100
C	110
D	101

2.8 OTHER USER-DEFINED MODELS
OF TRANSFORMATION

Although the tendency in cladistic studies has been to minimize assumptions
built into analyses, increased levels of sophistication with regard to computer
software have led to many 'user-defined' options being available for analysis,
especially in PAUP v.3.0. A brief mention of these 'user-defined' options will
suffice to introduce two of them. As described in Chapter 1 (Fig. 1.6a,b) certain
cladogram topologies could be compatible with equally parsimonious, but
conflicting, character transformations. The two alternatives in which a transformation is explained either as a parallel gain or as a synapomorphy with
secondary loss are reconsidered here (Fig. 2.5a,b). If, prior to analysis, it is
decided that characters should take the form of secondary loss rather than
parallel gain, then this hypothesis of character transformation can be included
as an integral part of analysis. The Dollo parsimony option in PAUP v.3.0
prohibits parallel gain (Fig. 2.5b) and stipulates that a character must only occur
once on the cladogram and therefore homoplasy always takes the form of
secondary loss.

The default setting for transformations between character states of a multistate
character are treated as having equal cost in terms of the number of steps. A step
matrix enables a separate value to be given to each step of a linear or branched
multistate character. This practice of weighting some transformations over
others is used in molecular systematics for weighting transversions over transitions as the latter occur more frequently (see also Chapters 3 and 7).

2.9 CHARACTER POLARITY

Returning to the example given in Table 2.1, species which have either two or
four stamens can be coded as a binary character. In its most general form this
coding allows the transformation to proceed in either of two directions, from
$0 \rightarrow 1$ or $1 \rightarrow 0$. If the direction of the transformation is determined, then the
character is said to be polarized and one or other of the two possibilities would

Fig. 2.5 User-defined models of character transformation. The two interpretations are equally parsimonious. (a) Character distribution interpreted as
parallel gain. (b) Prohibition of parallel gain; character only occurs once on a
cladogram, homoplasy explained as secondary loss.

be chosen for analysis. The determination of character polarity is particularly important for determining plesiomorphic and apomorphic characters.

The polarity of multistate characters determines the direction of transformation but not the order of transformation. Take the example in Fig. 2.3 which shows an ordered multistate character with three states. To polarize this ordered character is to determine the initial starting point of the transformation. Consequently, the ordered character sequence 0, 1, and 2 can be polarized in any of three ways, as shown in Fig. 2.6a–c. The multistate character with three states can be ordered in three ways and each of these can be polarized in three different ways, giving nine possibilities to constrain multistate characters with three states.

It is possible to polarize an unordered multistate character. (Given a binary character 0 − 1, with a specified direction of transformation, then whether the character is said to be ordered or polarized is a moot point.) Given an unordered multistate character which has three character states 0 − 1 − 2, it is possible to polarize the character in three ways, by choosing 0, 1, or 2 as the starting point of the transformation, and leaving the other states unordered. This then results in three possibilities for any one polarity decision as shown in Fig. 2.7a–c.

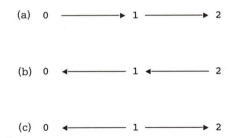

Fig. 2.6 Order and character polarity. The ordered character sequence 0, 1 and 2 may be polarized in any of the following three ways. (a) (0) most plesiomorphic. (b) (2) most plesiomorphic. (c) (1) most plesiomorphic.

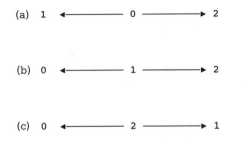

Fig. 2.7 Three possibilities for polarizing an unordered, 3 state character. (a) (0) most plesiomorphic. (b) (1) most plesiomorphic. (c) (2) most plesiomorphic.

2.10 CLADOGRAMS AND ROOTS

The results from analysing a data matrix can either be in the form of an unrooted tree (network) or a rooted tree (cladogram). Figure 2.8a−c shows all three possible unrooted solutions for four taxa. To root an unrooted tree involves imparting polarity onto at least one character transformation, although it is often the case that all or many characters will be polarized. For four taxa there are 15 possible fully resolved cladograms (Fig. 2.9).

The usual method for rooting a cladogram is outgroup comparison (see Chapters 3 and 6). This method involves choosing the sister group (or another closely related taxon) of the study group to root the cladogram. Rooting a cladogram determines the monophyletic groups, reveals paraphyly, and discovers relatively apomorphic and plesiomorphic characters.

By way of an example, Table 2.6 shows a data matrix for four taxa each with four binary characters. The study group is comprised of taxa 2, 3, and 4, with taxon 1 as the sister group. The cladogram shown in Fig. 2.10a is the most parsimonious for these taxa. The choice of root (outgroup), polarizes the characters within the ingroup and determines which states of the binary characters are apomorphic or plesiomorphic. Rooting on zeros also has the effect of grouping taxa solely on the presence of characters.

Figure 2.10a shows that characters A(1), B(1), and C(1) are synapomorphic and D(1) autapomorphic. The cladogram also determines that (3 4) and (2 3 4) are monophyletic groups. Consider that there may be some disagreement about the initial ingroup status of (2 3 4) (ingroups should be monophyletic) and that an alternative interpretation is for (1 2 3) to be monophyletic and 4 the sister group (Fig. 2.10b). This shows that A(0), B(0), and D(0) are synapomorphic and C(0) is autapomorphic. Figure 2.10b also shows that group (1 2 3) is monophyletic. This example illustrates the crucial point that even give the same data matrix the choice of root for the cladogram is of vital importance as it determines the status of cladograms and groups.

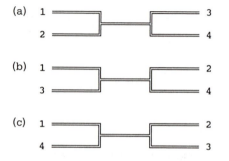

Fig. 2.8 All three possible unrooted solutions for four taxa.

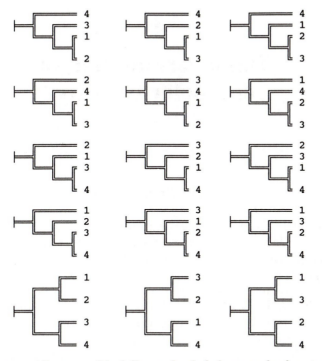

Fig. 2.9 All 15 possible fully resolved cladograms for four taxa.

Table 2.6 Binary coding for four taxa (1–4) and four characters (A–D); data for Fig. 2.10

Taxa	ABCD
1	0000
2	0010
3	1110
4	1111

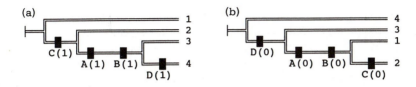

Fig. 2.10 Rooting, apomorphy and monophyly. (a) The most parsimonious tree taking taxon 1 as the root; characters A(1), B(1) and C(1) are synapomorphic and D(1) autapomorphic. (b) An alternative interpretation rooted on taxon 4: taxon (1 2 3) is monophyletic; and characters A(0), B(0) and D(0) are synapomorphic and C(0) is autapomorphic.

3.

The determination of character polarity

Ian J. Kitching

3.1 INTRODUCTION

Swofford (1990) distinguished three properties of characters: direction, order, and polarity. Unlike the definition of Meacham (1984) in which directed characters are equivalent to polarized characters, Swofford's definition of direction refers to 'cost' in terms of tree length of a change between any two character states. An undirected character is one in which the costs are symmetrical; that is, the increase in tree length required by the transformation of state X to state Y is the same as that required by the change from Y to X. An example of a directed character is one that is optimized under the Camin–Sokal parsimony criterion, where reversals to a more plesiomorphic condition are not permitted. Step matrix characters (Swofford 1990) are directed if the step matrix is asymmetrical (see also Chapter 4). Character order specifies the type of the permitted character state transformations and has already been introduced in Chapter 2.

Polarity refers to the direction of character evolution. A character is said to be polarized if the state ancestral to all other states is prespecified. Many methods and criteria for assessing the evolutionary polarity of characters have been proposed, and have been reviewed by Crisci and Steussy (1980), de Jong (1980), Stevens (1980), and Arnold (1981). Of these, most can be reduced to variations of three main themes: the outgroup comparison (the indirect method), the ontogenetic criterion (the direct method), and criteria based on specific models.

Nelson (1973a) divided the criteria used to estimate ancestral character states into two approaches. Indirect arguments (the indirect method) involve taxa other than those of the study group and rely upon a pre-existing higher-level phylogeny to establish character polarity. Farris *et al.* (1970) demonstrated the fundamental importance of parsimony to outgroup comparison, which was therefore considered by Nelson (1978) to be the only valid indirect method. However, the higher-level phylogeny itself must be based upon yet more polarized characters, leading to an infinite regress. Eventually, a method independent of pre-existing phylogenetic hypotheses must be invoked in order to validate the outgroup comparisons (Weston 1988). Of such direct arguments (the direct method), only ontogenetic character precedence was considered to be valid by Nelson (1973*a*).

3.2 OUTGROUP COMPARISON—THE INDIRECT METHOD

In its simplest form, outgroup comparison has been defined by Watrous and Wheeler (1981, p. 5) as:

For a given character with 2 or more states within a group, the state occurring in related groups is assumed to be the plesiomorphic state.

Watrous and Wheeler (1981) then gave a series of operational rules for a procedure they called the functional ingroup/functional outgroup (FIG/FOG) method. An initial hypothesis of relationships of outgroups and an unresolved ingroup is chosen. Characters are then selected and polarized using outgroup comparison, allowing partial resolution of the ingroup. Functional outgroups and functional ingroups are then established, which then permit further resolution. The procedure is repeated until either the ingroup is fully resolved or no further resolution can be achieved.

The following example, adapted from Mooi (1989), is based on the hypothetical electrophoretic allele data of Swofford and Berlocher (1987) (Table 3.1) and is optimized using the Fitch (1971) parsimony criterion (see also Chapter 4). Taxon F of the ingroup is established as the functional outgroup to the remaining ingroup taxa, A–E (Fig. 3.1a), by the presence in F and the outgroup, G, of allele 1c. Clade A–E is thus characterized by either allele 1a or 1b, but at this point in the analysis, a decision cannot be made between them.

Table 3.1 Hypothetical allelic data for seven taxa. (Adapted from Swofford and Berlocher 1987)

Taxa	Characters			
	1	2	3	4
A	b	c	a	c
B	b	c	a	c
C	b	c	b	a
D	a	b	b	a
E	a	b	b	a
F	c	d	a	a
G	c	a	c	b

Furthermore, because the alleles present in G for characters 2–4 may be autapomorphic for G or plesiomorphic for all taxa, polarity determination for these must wait.

The second iteration uses F as the FOG to A–E, ignoring the outgroup G. Alleles 4c and 3b are seen to be synapomorphic for taxa A + B and C + D + E respectively (Fig. 3.1b). Note that no decision can yet be made regarding the ambiguous status of alleles 1a and 1b.

The third iteration uses taxa A and B as the FOG to C + D + E, resulting in D + E being recognized as a clade, based upon alleles 1a and 2b. The initial ambiguity is also resolved: allele 1b is interpreted as synapomorphic for taxa A–E, with a subsequent transformation to allele 1a in taxon D + E (Fig. 3.1c). The final cladogram (Fig. 3.1d) shows one possible maximally parsimonious reconstruction (MPR; see also Chapter 4), in which alleles 1c, 2d, 3a, and 4a are interpreted as plesiomorphic for the ingroup + outgroup. Other MPRs exist; for example, alleles 2a, 3c, and 4b may be treated as plesiomorphic, whence alleles 2d, 3a, and 4a will unite taxon F with the remaining ingroup taxa as a monophyletic group.

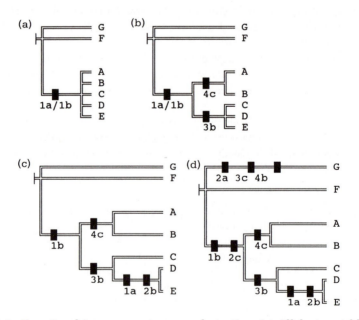

Fig. 3.1 Functional ingroup–outgroup polarization. (a) Allele 1c establishes F as the functional outgroup to the functional ingroup, A–E, which is characterized by either allele 1a or 1b. (b) With F as the functional outgroup, alleles 3b and 4c characterize (C + D + E) and (A + B) respectively. (c) Using A + B as the functional outgroup of C + D + E, D + E is recognized as a clade based on alleles 1a and 2b. Allele 1b is then seen to characterize (A–E). (d) Final resolution.

While Watrous and Wheeler's (1981) formulation is adequate when there is no variation in state within the outgroups, there are severe difficulties in applying this rule consistently when the outgroup taxa are heterogeneous. A number of specific examples of such outgroup variation have been examined by Arnold (1981) and Farris (1982*a*), and a comprehensive overview is provided by Maddison *et al.* (1984).

Maddison *et al.* (1984) presented a general algorithm and principles to enable the most parsimonious hypothesis of an ancestral state to be estimated given fully resolved and fixed outgroup interrelationships, and then discussed the effects of uncertainty among those relationships. Such examination of outgroups ensures that the ingroup cladograms are globally parsimonious. If the ingroup alone is studied, as in the the procedure of commonality (see below), then the chosen ingroup cladogram will only be locally parsimonious. Failure to achieve global parsimony may also result if outgroup analysis is taken to indicate the state present in the most recent common ancestor of the ingroup (for example Wiley 1981) or if the ingroup is first resolved in isolation as an unrooted tree without reference to ancestral states and the outgroup subsequently attached, as in the procedure of 'Lundberg rooting' (Lundberg 1972; Swofford 1990). Similarly, if the state occurring most commonly among the outgroups is assumed to be plesiomorphic (Arnold 1981), then non-globally parsimonious solutions may result, depending upon the precise distribution of the character state and the relationships of the outgroup taxa.

Several terms and conventions were defined by Maddison *et al.* (1984) to assist their general discussion. The most recent common ancestor of the ingroup taxa is called the ingroup node, while the next most distal node, which links the ingroup to the first outgroup is the outgroup node (Fig. 3.2). Furthermore, they assumed that the outgroup interrelationships were known and immutable.

The method of Maddison *et al.* (1984) aims to estimate the character state of the outgroup node. Such an assignment may be either 'decisive', if it can have only a single value, or 'equivocal' if it can have more than one equally parsimonious alternative state. They also noted that their results hold whether cladograms are interpreted as indicating recency of common ancestry (Nelson 1973*c*) or a pattern of nested sets of characters (Nelson and Platnick 1981).

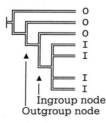

Ingroup node
Outgroup node

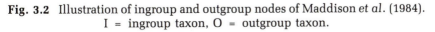

Fig. 3.2 Illustration of ingroup and outgroup nodes of Maddison *et al.* (1984).
I = ingroup taxon, O = outgroup taxon.

Simple examples can be determined by visual inspection. For example, if there is only one outgroup with state x, it is more parsimonious to assign state x to the outgroup node than to assign some other state (Fig. 3.3a; but see below). Such an assignment is decisive. If the first two outgroups (Fig. 3.3b) differ in their states, then the assignment to the outgroup node is equivocal. Additional, more distal, outgroups with the same state as the second outgroup exert no influence, demonstrating that the assignment is not simply a function of the relative frequency of the states within the outgroups (but see Arnold 1981).

However, such visual inspections may fail to achieve the most parsimonious assignment in more complex cases of heterogeneous outgroups and an algorithmic approach is necessary. Maddison *et al.* (1984) adapted the methods used in Wagner and Fitch optimization to outgroup analysis. These methods are discussed in greater detail in Chapter 4 and only the outgroup comparison using binary (x/y) characters will be considered here.

A general algorithm for multistate characters was given by Maddison *et al.* (1984). First, the terminal outgroup taxa are labelled with their observed states, x or y (Fig. 3.4). Polymorphic outgroups, if any, are labelled xy. Then, beginning with pairs of terminal outgroups and proceeding towards the outgroup node, the internal nodes are labelled according to the following rules:

1. If the derivative nodes are both labelled x, or are x and xy, the ancestral node is labelled x.

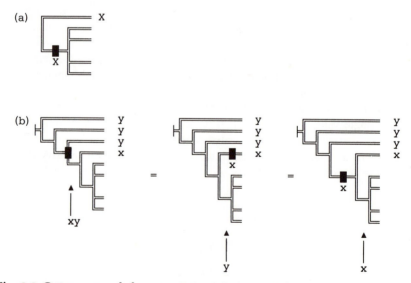

Fig. 3.3 Outgroups and character state assignment at the outgroup node. (a) If there is only a single outgroup, then the assignment of the character state of the outgroup to the outgroup node is decisive. (b) If there are multiple outgroups, then if the first two outgroups disagree in their state assignments, the character state assignment at the outgroup node is equivocal.

2. If the derivative nodes are both labelled y, or are y and xy, the ancestral node is labelled y.

3. If the derivative nodes are x and y, or both xy, the ancestral node is labelled xy.

Ignoring the root and proceeding towards the outgroup node simplifies the procedure by negating the need for a preorder traversal of the tree in order to determine the MPR (see also Chapter 4).

Two simple rules follow from this procedure that allow for quick and accurate ancestral character assessment.

1. *The first doublet rule* (Fig. 3.5a,b). If the first outgroup and the first doublet (a pair of consecutive outgroups that agree in a state) share the same state, then that state is the decisive maximally parsimonious reconstruction (MPR). If they disagree, then the decision is equivocal. Furthermore, if the first two outgroups form a doublet, then their state is the MPR. A corollary to this rule is that all outgroup structure beyond the first doublet is irrelevant to the assessment.

2. *The alternating outgroup rule* (Fig. 3.6a,b). If there are no doublets, then if the first and last outgroups agree, this state is the decisive MPR; otherwise the decision is equivocal. These rules are applied with no regard for the distribution or composition of the ingroup character states. For example, if states x and y are found in the ingroup, and states x and z in the outgroup, both de Jong (1980) and Watrous and Wheeler (1981) suggested that state x is the ancestral state for the ingroup, because this is the only state to occur in both ingroup and outgroup. However, Farris (1982*a*) demonstrated that ignoring state z in this way may sometimes lead to a non-parsimonious result.

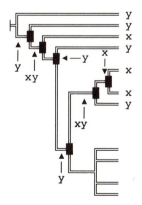

Fig. 3.4 Illustration of the algorithmic approach of assigning a character state to the outgroup node in a cladogram with heterogeneous outgroup terminals (Maddison *et al.* 1984).

Whether z, x or xz is the most parsimonious reconstruction will depend upon the outgroup relationships (Maddison *et al.* 1984).

A problem arises in the situation in which only one outgroup is used (Fig. 3.3a), for there is no outgroup node for which an MPR can be assessed (the apparent root of the tree is ignored when the general algorithm is applied). Maddison *et al.* (1984) implicitly adopted the convention that the single outgroup itself forms the outgroup node, and thus whatever state occurs in that outgroup is automatically considered to be plesiomorphic. But this convention is properly valid only if the tree is drawn as having no subtending basal branch. If the tree is so rooted (Fig. 3.7a), and the ingroup resolved (Fig. 3.7b), then

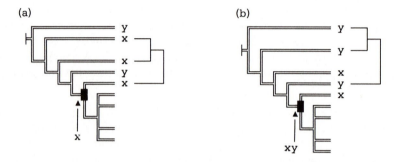

Fig. 3.5 Illustration of the first doublet rule for binary characters (see text). (a) If the character state of the first outgroup agrees with that of the first doublet, the character state assignment at the outgroup node is decisive. (b) If the character state of the first outgroup disagrees with that of the first doublet, the character state at the outgroup node is equivocal.

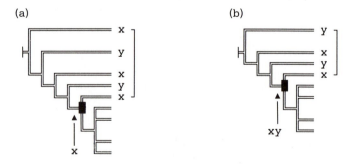

Fig. 3.6 Illustration of the alternating outgroup rule for binary characters. (a) If the character states of the first and last outgroups agree, the character state assignment at the outgroup node is decisive. (b) If the character states of the first and last outgroups disagree, the character state assignment at the outgroup node is equivocal.

a hidden assumption is involved regarding the state present at the base of the tree, which is that this state agrees with that in the sole outgroup (Fig. 3.7c). Then, by the first doublet rule, the state at the outgroup node is a decisive x, and y can be interpreted as apomorphic within the ingroup. Usually, however, no evidence is provided to support this assumption (for example Humphries and Funk 1984, fig. 5). There is an alternative assumption, which is that the base of the tree possesses state y (Fig. 3.8a). If so, then the outgroup node state is equivocal and y cannot be decisively assessed as apomorphic (Fig. 3.8b,c). Now it is true that the tree in Fig. 3.7c is more parsimonious than either of those in Figs. 3.8b and 3.8c, and could thus be chosen as the preferred interpretation on that basis alone. However, in order to avoid the hidden assumption, it is recommended that at least two outgroups be used in an analysis employing outgroup comparison.

The algorithm of Maddison *et al.* (1984) produces the globally most parsimonious ingroup cladograms, even though the ingroup character state distributions are not taken into account. The procedure is sufficient providing the sole aim is to resolve the ingroup relationships. However, Maddison *et al.* (1984)

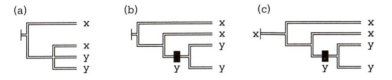

Fig. 3.7 Character state assignments for binary characters at the outgroup node for cladograms with a single outgroup. (a) Rooted cladogram with a single outgroup and character states assigned to terminals. (b) Cladogram of Fig. 3.1a with the ingroup resolved. (c) However, the resolution in Fig. 3.1b assumes that the state at the base of the cladogram is the same as that of the outgroup, hence the first doublet rule applies.

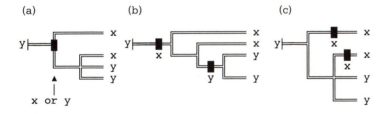

Fig. 3.8 Character state assignments for binary characters at the outgroup node for cladograms with a single outgroup. (a) Rooted cladogram with a single outgroup and assuming y is the basal state. Assignment at the outgroup node is equivocal (see text). (b) One possible MPR given the conditions in (a). (c) Alternative MPR given the conditions in (a).

noted that if the ancestral state assignments are required for other purposes, such as the evolutionary modelling of character transformation, then the ingroup should be resolved as far as possible first and then either Wagner or Fitch optimization applied to the entire tree.

Problems arise in using the algorithm when the outgroup relationships are uncertain or are not prespecified. Clearly, if all the outgroups agree in state, then their interrelationships are irrelevant. However, if they differ, then uncertainty in outgroup relationships can lead to uncertainty about the ancestral state. The algorithm could be applied to all possible outgroup resolutions, but this would be remarkably tedious for more than a very small number of outgroups. Maddison *et al.* (1984) gave six rules derived from the outgroup criterion that could be used to indicate the degree to which differing outgroup relationships may yield the same MPR. These rules describe situations in which outgroup uncertainties have no or limited effect. ('Limited' is defined as not allowing changes that will completely shift an assessment; that is (for a binary character), assessments can change from decisive to equivocal (for example x to xy, or xy to y) but not from decisive to decisive (x to y). For multistate characters, the alternative assessments must overlap, that is, they must both contain at least one state in common.)

1. Uncertainties beyond the first doublet have no effect.

2. If the root is moved within the outgroups, there is no effect on the MPR.

3. If the first outgroup or the basal node of the first subgroup of outgroups has one state, then the MPR is either decisive or equivocal for that state. The first outgroup can thus be seen to exert a considerable influence on the MPR. However, even though this taxon may be highly derived, there is no justification in appealing to more distant but supposedly more primitive outgroups.

4. The addition or deletion of one outgroup cannot completely shift an MPR; at least two additions or deletions are necessary to accomplish this. However, even the addition of a distant outgroup can affect the MPR (cf. Figs. 3.6a and 3.6b).

5. Similarly, moving one outgroup cannot completely shift an MPR. Maddison *et al.* (1984) demonstrated that this rule held for binary characters but were unable to prove or disprove it for multistate characters in general.

6. When only one of a number of outgroups possesses a particular state, the MPR cannot be decisive for that state whatever the outgroup relationships.

When there are alternative ancestral state assignments depending upon alternative outgroup resolutions, there are a number of means by which a cladistic analysis can proceed. The optimum approach would be first to conduct a higher level analysis to resolve the outgroup relationships fully. However, this rapidly leads to an infinite regression of ever higher-level analyses. In the absence of such analyses, the assessment of ancestral states could be determined under a more restrictive parsimony model (for example Dollo parsimony). Alternatively, an appeal could be made to the 'predominant states method', i.e. outgroup commonality (Arnold 1981). However, the last approach is not philosophically justified by a direct parsimony argument and may give non-globally parsimonious cladograms.

If full resolution of outgroup relationships is not feasible, then even partial resolution may reduce the number of uncertain MPRs. It may then be practicable to examine the influence of all outgroups on the MPR, both singly and in all allowable combinations (the outgroup substitution approach of Donoghue and Cantino 1984). The various ingroup resolutions could then be examined for areas of congruence using the strict consensus tree technique. Clades present in the consensus tree would be those unaffected by uncertainty in outgroup relationships. The ingroup could then be further resolved using the functional outgroup/ingroup (FIG/FOG) technique (Watrous and Wheeler 1981). Maddison *et al.* (1984) suggested resolving the ingroup cladogram using each of the possible ancestral states and selecting those cladograms that were most parsimonious according to the outgroups, ingroups or both.

The above technique can be characterized as a constrained, two-step analysis, in which the MPR of the outgroup node is first assessed, then followed by resolution of the ingroup. Clark and Curran (1986) argued that unconstrained, simultaneous analysis is superior. In this procedure, the most parsimonious cladogram for both outgroups and ingroups is estimated in one step, with no constraints placed upon the permitted resolution of either. Clark and Curran (1986) identified two a priori assumptions required by the constrained analysis of Maddison *et al.* (1984).

1. The ingroup is monophyletic, which implies that the root is basal to it.

2. Outgroup structure implies hypotheses of monophyly that are not open to testing because they are treated as immutable.

Problems can thus arise for characters that show parallel homoplasy between one or more outgroups and a subset of the ingroup. In Fig. 3.9a, where A−C are the ingroup and D−F the outgroups, character state 4 is synapomorphic for the group, A−E. A subsequent transformation to 4− characterizes the clade A + B. If two further characters are found with the same distribution as character 4, then either the outgroup formulation of Watrous and Wheeler (1981) or the general constrained algorithm of Maddison *et al.* (1984) will yield a most parsimonious tree of nine steps (Fig. 3.9b). However, if all six characters are analysed without topological constraints, then a very different, and shorter, cladogram of eight steps is found (Fig. 3.9c), in which the ingroup is not monophyletic. The lack of global parsimony in Fig. 3.9b is due to the requirement of immutability of outgroup relationships.

Clark and Curran (1986) and Farris (1982a) noted that a simultaneous, unconstrained analysis would never give a less parsimonious result than a two-step, constrained analysis. The method is fully consistent with Watrous and Wheeler's (1981) 'operational rule' for the outgroup criterion. Furthermore, it can be used where the outgroup relationships are unresolved; indeed, it may help to resolve them. Therefore, because simultaneous analysis makes the fewest a priori assumptions, it is the parsimony method of choice using outgroup comparison.

3.3 ONTOGENY — THE DIRECT METHOD

Within cladistics, there are two contrasting viewpoints regarding the role of ontogenetic information (Eldredge 1979; Williams *et al*. 1990). The transformational approach, which corresponds closely to Hennig's (1966) phylogenetic systematics, is performed in three stages (Weston 1988): character transformation series are formulated, which are then polarized and used to

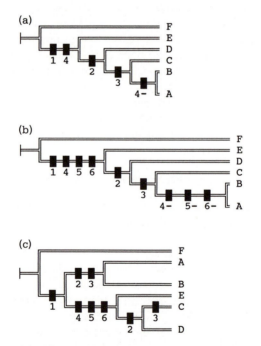

Fig. 3.9 Outgroup-constrained and outgroup-unconstrained analysis. Taxa A–C comprise the ingroup and D–F the outgroups. (a) Cladogram with characters assigned to nodes and outgroup topology predetermined and constrained. (b) Addition of two further characters (5 and 6) with the same state distribution as character 4 gives a cladogram of 9 steps with the same topology as that in Fig. 3.9a. (c) However, if the constraint on outgroup relationships is removed, a shorter cladogram of eight steps is possible in which the ingroup is not monophyletic.

construct cladograms. This contrasts with the taxic approach, considered to be embodied within pattern cladistics (Beatty 1982), which uses the distribution of homologies to hypothesize group membership. Here, character polarity is derived from the analysis rather than being an a priori assumption (Nelson and Platnick 1981; Patterson 1982*a*; Nelson 1985). A comparison of these two approaches raises the question of whether it is at all possible to determine character polarity prior to analysis.

Nelson (1978) generalized his direct argument thus:

Given an ontogenetic character transformation, from a character observed to be more general to a character observed to be less general, the more general character is primitive and the less general character is advanced.

He explained it using the following example (Fig. 3.10a). Suppose there are two taxa, A and B, possessing characters x and y respectively. With this information alone, no decision can be made regarding which character is the more primitive. However, a study of the ontogeny of the two species reveals that embryos of both species have character x but that during the subsequent development of species B, character x transforms into character y. In other words, x is observed to be more general and y to be less general. Character x is therefore inferred to be plesiomorphic and character y apomorphic. Both Nelson and Platnick (1981) and Patterson (1982*a*) considered that this reformulation of the biogenetic law was the decisive criterion in determining character polarity.

However, Lundberg (1973) argued that if the ontogenetic transformation is treated as the character, then both cladograms are equally parsimonious (Fig. 3.10b). The essence of his argument is as follows: if the transformation of x → y

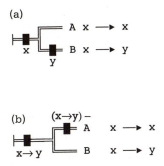

Fig. 3.10 (a) Nelson's (1973*b*) example of polarization using the direct ontogenetic criterion; state x, occurring in the ontogeny of both species, is the more general state and is therefore plesiomorphic. (b) Lundberg's (1973) counter-example. The gain of the transformation of x → y is regarded as a single step at the base of the cladogram, rather than two steps (gain of x, followed by gain of y). The subsequent loss of this transformation in taxon A gives a cladogram of two steps that is equally parsimonious as the cladogram in Fig. 3.10a.

is regarded as a single character, rather than as two characters, then the primitive gain of x → y and the subsequent loss of y only involves two steps, not three, as Nelson's reasoning would have it. This is then an equally parsimonious alternative to the primitive gain of x and its subsequent transformation during ontogeny into y. Nelson (1973b) suggested that this is a question of negative gains or losses which is contrary to a simple parsimony argument. Nelson (1978) rebutted Lundberg's interpretation, demonstrating that the reformulated biogenetic law was logically justified by parsimony.

Essentially similar arguments were subsequently put forward by Alberch (1985), Kluge (1985), and De Queiroz (1985). In particular, De Queiroz objected to the use of 'instantaneous morphologies' (= horizontal characters, Kluge 1988a) as characters, because they are abstractions from 'real' ontogenetic transformations. He adopted the viewpoint that phylogeny is a sequence of life cycles and considered that the evidential basis for inferring historical relationships ought to be the ontogenetic transformations themselves, not the features that are transformed (Kluge 1988a). As a result, he concluded that there could be no 'ontogenetic method'.

Kluge (1988a) also examined De Queiroz' (1985) concept of treating transformations as characters but concluded that it was incomplete. Furthermore, it offered no unique advantage over describing the life cycle in terms of the model of growth and differentiation developed by Fink (1982) and Kluge and Strauss (1985) from the work of Alberch *et al.* (1979) and Alberch (1985).

Taken to the extreme, De Queiroz' approach would view the entire organism, if not the whole of the living world, as a single character and there would be no basis for comparative biology (Weston 1988). De Queiroz (1985) actually adopted a more pragmatic and reductionary approach, defining 'character' as features of organisms that are 'large enough to encompass variation that is potentially informative about the relationships among the organisms being studied.' However, this approach is fraught with difficulties, not the least of which is that it is open to the same criticism as that of Lundberg (1973), for if 'large enough to encompass variation' is replaced by 'small enough to differentiate variation', the definitions become identical (Weston 1988).

Further controversy regarding Nelson's reformulation of the biogenetic law revolves around the precise meaning to be attached to the term 'general', of which there are two possible interpretations (Kluge 1985):

1. Strict temporal precedence, so that the more general state is that which arises first in ontogeny, the interpretation adopted by Rosen (1982) and Patterson (1982b); or
2. Simply commonness, regardless of ontogenetic sequence, the interpretation of Nelson (1978).

Kluge (1985) argued that because Nelson's law uses commonness as the estimator of polarity, it is just a special case of ingroup commonality, with all the failings associated with that technique (see below). However, De Queiroz

(1985) and Weston (1988) chose to interpret 'more general' as meaning not simply just 'more common'. They emphasized that although the more general character will be more common, because it is possessed by all those taxa that also possess the less general character as well as some taxa that do not, this does not equate with ingroup commonality. Commonality bears no necessary relationship to relative time of phylogenetic appearance and thus there is no reason to infer that common characters are ancestral to less common ones. A character may be more common than another but unless there is an unequivocal relationship of generality, it will not be more general (Weston 1988). Furthermore, equating generality to strict temporal precedence results in Haeckel's biogenetic law. This has long been refuted, because it can only operate correctly if ontogenetic change occurs solely by terminal addition (Kraus 1988).

In fact, there are six fundamental ways in which ontogenetic sequences can change: additions, substitutions, and deletions, which can be terminal or non-terminal (Fig. 3.11; O'Grady 1985) (Mabee (1989) added a seventh: substitution of an entire sequence, but this is just an extreme case of non-terminal substitution.) However, an important feature of the examples used by Nelson (1973*a*, 1978) is that the characters used are 'epigenetic' (Løvtrup 1978). For such characters, each ontogenetic stage is a modification of, or is induced by, a pre-existing ontogenetic stage and thus is developmentally dependent upon its precursor. If this dependence remains unaltered by subsequent evolution, the sequence of appearance cannot change. It follows that, in such characters, new states can be terminally added or deleted, or substituted non-terminally, but cannot be added or deleted non-terminally. Nor can the sequence be scrambled. However, non-terminal substitutions result in information loss, necessitating the use of outgroup comparison to resolve the transformation series. Terminal deletions give secondarily simplified ontogenies that Nelson's ontogenetic criterion cannot distinguish from primitive, unmodified ontogenies. Nelson (1973*a*, 1978, 1985) argued that such errors would be detected by comparative analysis of other characters and the application of parsimony. Nevertheless, using the epigenetic model for ontogenetic character transformation, Nelson's biogenetic law can only analyse characters in which ontogenetic change has occurred by terminal addition (Brooks and Wiley 1985; Kluge 1985; Kraus 1988; Weston 1988).

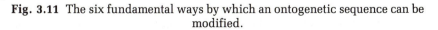

Fig. 3.11 The six fundamental ways by which an ontogenetic sequence can be modified.

Brooks and Wiley (1985) examined the relative merits of the ontogenetic criterion and outgroup comparison. They concluded that of the six classes of ontogenetic change, the two criteria would only give the same result for terminal and non-terminal additions. In all other cases, the ontogenetic criterion would be misled by paedomorphosis. In order to resolve characters in which paedomorphosis had occurred, the outgroup criterion had to be implemented (Eldredge and Cracraft 1980; Kluge 1985; O'Grady 1985). Thus, because direct observation of ontogeny did not resolve any instances of evolutionary change that could not also be resolved by use of outgroup comparison, and because outgroup comparison did resolve other cases that ontogeny failed to resolve, Brooks and Wiley (1985) viewed the former as an incomplete method of character polarization relative to the latter.

Kluge (1985) also investigated the claims of Nelson and Platnick (1981) and Patterson (1982a) that the ontogenetic criterion was without theoretical pre-judgment. He concluded that, like outgroup comparison, the ontogenetic criterion was not theory-neutral and that its use in deducing sister group relationships required certain assumptions. Furthermore, to adopt the position that the direct observation of ontogeny is both necessary and sufficient to perform phylogenetic analysis would 'actually prevent discovery of phenomena such as paedomorphosis and non-linear ontogenetic sequences' (Kluge 1985).

Nelson (1985) responded to these criticisms, examining each of the examples used by both Brooks and Wiley (1985) and Kluge (1985) in turn. He demonstrated that Brooks and Wiley (1985) were modelling evolutionary not ontogenetic changes. Furthermore, their assertions that outgroup comparison resolved instances of terminal and non-terminal deletion were erroneous, while 'their terminal and non-terminal substitutions are "correctly" resolved only by ignoring that the data, independent of ontogenetic transformation, specify most parsimonious resolutions that differ from their "correct" results but agree with ontogeny' (Nelson 1985). The examples of Kluge (1985) were likewise shown not to support his conclusions. As for the phenomenon of paedomorphosis, Nelson (1985) considered it to be a problem for systematics in general and not ontogeny in particular. However, it remains true that any character where ontogenetic change proceeds from the less general to the more general would be ignored for the purposes of phylogenetic analysis under Nelson's ontogeny criterion (Kraus 1988).

Weston (1988) recognized this limitation of Nelson's biogenetic law. However, he showed that it was possible to hypothesize situations that did not conform to the requirement for terminal addition, but which nonetheless could still be analysed using a more broadly framed direct method of character analysis. The ontogeny shown in Fig. 3.12a (Rosen 1982) cannot be analysed under Nelson's biogenetic law because the epigenetic interactions have changed during the course of their evolution. This is indicated by the most parsimonious solution implying non-terminal additions. Such an approach, however, is one of 'constrained parsimony' (Nelson 1985), in which the constraint is that the cladograms minimally violate the ontogenetic transformations. If this constraint is removed, an alternative topology is obtained (Fig. 3.12b). However, parsimony cannot be invoked to choose between the two topologies (Fig. 3.12a,b) because each is most parsimonious for its own data set (Nelson 1985). The choice depends upon whether or not the constraint is justified, which is open to question (Kluge 1985; Nelson 1985; Weston 1988).

Weston (1988) also considered the analysis of serially homologous characters, which are essentially ontogenetic 'bushes' in which there is little or no epigenetic interaction between the branches. Thus some branches can change while others remain unaltered (Fig. 3.13). He also examined organogenetic sequences and

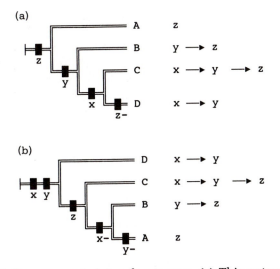

Fig. 3.12 Cladogram constraint and ontogeny. (a) This ontogeny cannot be analysed under Nelson's reformulated biogenetic law because the epigenetic interactions have changed during the evolution of taxa A–D. (b) If the constraint that cladograms should violate minimally the observed ontogenetic transformations is removed, then an alternative topology is obtained.

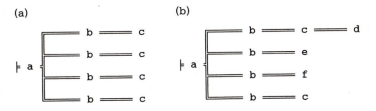

Fig. 3.13 Ontogenetic sequences in serial homologues. (a) Initial common ontogenetic sequence. (b) Because there are no epigenetic interactions among branches, subsequent modification in different serial homologues allows some to change while others remain unaltered.

biosynthetic pathways, concluding that the non-uniqueness of such sequences was irrelevant to the application of the direct parsimony argument. The use of such characters greatly increases the scope of application of the direct argument. This is of particular importance for work at very high cladistic levels (for example Lipscomb 1985), where the lack of meaningful, or even any, outgroups can be extremely problematic for the indirect method.

By emphasizing the role of directly observed generality relationships between homologous characters, Weston sought to remove any reference to sequence in the direct method. He thus was able to further generalize Nelson's law:

Given a distribution of two homologous characters in which one, x, is possessed by all of the species that also possess its homolog, character y, and by at least one other species that does not, then y may be postulated to be apomorphous relative to x.

However, to ignore this orderliness involves the potential loss of some phylogenetic information. Nelson (1985) considered that data 'constrained by ontogeny are, in a way, the more complete'. Q. D. Wheeler (1990) agreed that the most objective application of parsimony to the problem of ontogenetic sequences must be sought but the use of ontogenetic data devoid of sequence information is not unequivocal. He therefore concluded that Nelson's original formulation was preferable to that of Weston. In addition, Wheeler investigated the potential influence of the four situations identified by Kraus (1988) that restricted the application of outgroup comparison:

(1) difficulties in determining closest relatives;

(2) difficulties in assessing homologies when the ingroup and outgroup were widely divergent;

(3) reduction in confidence of polarity estimates with increasing 'distance' between ingroup and outgroups; and

(4) ambiguous assessments of polarity resulting from poorly or unresolved outgroup relationships.

Wheeler found that all of these problems were real but also that they all applied equally to both Nelson's law and outgroup comparison. Wheeler nevertheless advocated the use of ontogenetic data, based not upon any fundamental difference between outgroup comparison and Nelson's law, but because the latter is founded upon the uniqueness of pairwise comparisons and the ability directly to observe ontogenetic adjacency (Nixon and Wheeler, in preparation).

Kluge (1985) also considered the problem of characters that do not demonstrate an obvious ontogeny. Some proteins (Uy and Wold 1977; Stryer 1981) do exhibit ontogenetic changes, but most do not. Similarly, most nucleotide bases and karyotypic characters also seem to be without ontogeny. Kluge (1985) also considered that there may be a similar problem when analysing the relationships of unicellular organisms that may also lack an ontogeny. Nelson

(1985) side-stepped these issues, stating that if the aim of the exercise was to investigate the relevance of ontogeny, then features and taxa without ontogeny were irrelevant.

The issue of whether unicellar organisms have ontogenies was addressed by Blackmore (1986). He concluded that all organisms must have ontogenies but because the processes operating at the cellular level are often poorly understood, the application of the ontogenetic criterion under such circumstances may be problematical. An example of the application of the ontogenetic criterion to unicellular organisms, diatoms, was provided by Kociolek and Williams (1987). However, the problem of the applicability of the ontogenetic criterion to nucleotide and similar characters remains (see Patterson 1988a).

For all the theoretical discussion about the relative merits of outgroup comparison and the ontogenetic criterion, there have been relatively few empirical studies. Miyazaki and Mickevich (1982) compared a cladogram for *Chesapecten* clams, derived using transformation series analysis without reference to developmental information, to observed ontogenetic changes. They concluded that study of the ontogenetic sequence of a character within an individual taxon provides polarity information and that Nelson's biogenetic law applied to their example. Kraus (1988) analysed a data set of 32 characters derived from *Ambystoma* salamanders, applying the ontogenetic criterion and outgroup comparison independently (using the methods of both Maddison et al. (1984) and Clark and Curran (1986)). Comparing the resulting cladograms for topological similarity, level of resolution, and degree of ambiguity, he concluded that the ontogenetic criterion performed as well as or better than the outgroup criteria (although the differences were small). Q. D. Wheeler (1990) performed a similar comparison using data derived from a group of staphylinoid beetles. He concurred with Kraus (1988) that the ontogenetic criterion is a viable alternative to outgroup comparison for polarizing characters. But whereas Kraus considered that, on the whole, the ontogenetic criterion outperformed outgroup comparison, Wheeler found that both methods produced similar numbers of equally parsimonious cladograms, which implied similar levels of homoplasy. Thus, neither method was considered to be superior to the other, although the fact that the ontogenetic criterion could be applied to a minimum of two taxa, rather than three as required by the outgroup criterion, was considered to be a possible advantage under certain circumstances.

In contrast, Mabee (1989), in a study that estimated the effectiveness of the ontogenetic criterion independently of outgroup hypotheses, concluded that although ontogeny could provide useful data, the ontogenetic criterion was not empirically justified as a method of inferring phylogenetic polarity. It is considered here that her argument is flawed because she assumes that the trees found by outgroup comparison are the 'correct' trees.

A general consensus seems to be emerging regarding the interrelationship between and the relative merits of direct and indirect criteria. Q. D. Wheeler (1990) concurred with Kraus (1988) that Nelson's reformulation of the biogenetic law is a viable alternative to outgroup comparison as a method for polarizing characters and that there are valid theoretical reasons to accept each. In fact, it is possible that they are not independent and that there is in fact no clear division between the two concepts (Nelson, cited in Q. D. Wheeler 1990). Both the ontogenetic criterion and outgroup comparison are valid approaches, because both are ultimately justified by parsimony (Farris 1983; Weston 1988). Some authors think that systematic analysis should involve both methods (Stevens 1980; Fink 1982; Blackmore 1986), although there are occasions on which only one or other criterion can be employed. When both methods are applicable, each can act as a countercheck on the other. In an ideal world, Nelson's law and outgroup comparison would give the same result. It is homoplasy that causes either method to fail (Q. D. Wheeler 1990).

3.4 INADEQUATE CRITERIA

In addition to outgroup comparison and ontogenetic precedence, numerous other methods for assessing character polarity have been proposed (for reviews see Crisci and Steussy 1980; de Jong 1980; Stevens 1980; Arnold 1981). Many of these criteria have been shown to be based upon unjustified or false assumptions and the majority can simply be ignored. However, two have been more widely referred to in the literature and need to be considered in greater detail: ingroup commonality, which is described below, and the stratigraphic criterion, which is discussed in detail in Chapter 8.

3.4.1 Ingroup commonality

Ingroup commonality can be defined as follows:

A plesiomorphous character state is more likely to be widespread within a monophyletic taxon than is any one apomorphous character state (de Jong 1980).

Or put simply − common is primitive. In this form, the criterion is *ad hoc* because it assumes that the evolutionary process tends to conserve plesiomorphic character states (Wiley 1981).

Although the commonest character state will be plesiomorphic in some groups of taxa, in other groups it will not. But it is impossible using ingroup commonality alone to distinguish those two instances. Only recourse to outgroup comparison or the ontogenetic criterion can resolve the problem, when the use of commonality becomes redundant. Furthermore, ingroup commonality cannot resolve the three-taxon problem (Watrous and Wheeler 1981; Humphries and Funk 1984), for to do so requires that two of the three taxa share an apomorphy, a character state distribution forbidden under commonality. In addition, commonality tends to produce symmetrical cladograms, while initially incorrect hypotheses of relationship will continue to be supported by additional characters if these are polarized using commonality alone. The ingroup commonality criterion is fallacious (Hennig 1966; Nelson and Platnick 1981; Watrous and Wheeler 1981) and should not be used to assess character polarity.

3.5 A PRIORI MODELS

There is a third group of criteria for polarity determination which depend for their justification upon the adoption of specific underlying models. Their validity is dependent upon the strength of these models, which can be divided into two types:

1. Models based upon specific hypotheses of the mechanism of evolutionary change. Consequently, these criteria may be of very restricted practical application. Examples include criteria based upon Robertsonian translocations and dominance hierarchies among alleles.

2. Vaguely formulated, contradictory or *ad hoc* models that are unjustified and fallacious, and that often prejudge the results of an analysis; for example the a priori use of 'function' and underlying synapomorphy.

3.5.1 Robertsonian changes

This type of chromosomal rearrangement involves the fusion of a pair of acrocentric or telocentric chromosomes to form a single metacentric chromosome. Jones (1974) considered that such a fusion required, in the case of acrocentric chromosomes, two asymmetrical breaks, one in the short arm of one chromosome and one in the long arm of the other. This would permit the resultant fragments to fuse into one long-armed metacentric and one extremely short-armed metacentric chromosome. The latter, being unable to form chiasmata dependably, has an extremely high probability of being lost during meiosis. The process is thus essentially irreversible. Using this criterion, Jones (1974, 1977) constructed a hypothetical scheme for the evolution of the chromosomes of the genus *Cymbispatha* (Commelinaceae).

3.5.2 Allelic dominance hierarchy

In a diploid organism, an allele is much more likely to become fixed in a population if it is dominant. Thus, Turner (1983, 1984) argued that, given a pair of alleles, the recessive is likely to be the plesiomorphic state. The dominance relationships among a set of alleles can therefore be used to determine both the sequence and direction of evolutionary change. He applied this criterion to the mimetic patterns of two species of *Heliconius* butterflies and was able to demonstrate that not only are these species extremely good parallel polytypic Müllerian mimics at the present time, but that they have been so throughout the entire period during which the mimicry was evolving. A similar approach was applied to the evolution of the polymorphic mimetic forms of the African swallowtail butterfly, *Papilio dardanus* (Vane-Wright 1979; Clarke 1980).

3.5.3 Functional morphology

It has been considered that an assessment of function is of great, even paramount, importance in the assignment of evolutionary character polarity (for example de Jong 1980; Bishop 1982). However, many such authors misuse the concept of function (Lauder 1990), using it as a synonym of 'selective value'. Such usage is fraught with difficulties (Davis and Heywood 1963; Kluge and Farris 1969; Stevens 1980; Lauder 1986), not the least of which is the need to accurately understand the nature of the selective forces acting on the structure (Lauder 1990). Most studies that purport to be 'functional analyses' simply present morphological data and then infer rather than measure function.

The correct interpretation of functional data is that in which the use of structural features has been directly observed and measured (Lauder 1990) and includes such characters as enzyme reaction rates and patterns of electrical activity during animal movement. Such functional characters are thus no different from any other character, being potential synapomorphies to be utilized in a cladistic analysis. Evolutionary polarity can then be inferred using the ontogenetic criterion and/or outgroup comparison.

A number of authors have argued that functional analysis provides an a priori guide to homology (for example Hammond 1979; Tyler 1988), independent of the distribution of characters in other taxa. This approach suffers from two major difficulties. First, as noted above, function is rarely directly measured, reliance being placed upon the heuristic use of functional ideas and hypothesized functions (Lauder 1990). Secondly, characters determined as homologous by a functional analysis are not subject to refutation by the discovery of another better-corroborated cladogram.

Hence, true functional characters can be admitted to a cladistic analysis but functional considerations should not be utilized in polarity determination.

3.5.4 Underlying synapomorphy

Underlying synapomorphy is defined as 'close parallelism as a result of common inherited genetic factors causing incomplete synapomorphy' (Saether 1983) or 'the inherited capacity to develop parallel similarities' (Saether 1986). Consequently, the apomorphic state is only observed in some members of a monophyletic group. In Fig. 3.14, of the study group A–H, taxa B, C, and H show the putative apomorphic state y. Saether (1979, 1983, 1986) would infer from this distribution that the ancestor of the group A–H had developed an underlying synapomorphy, (y*), which thereby demonstrates monophyly of A–H. However, this feature is not the observed expression of the state y, but

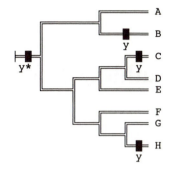

Fig. 3.14 The concept of underlying synapomorphy suggests that character y implies support (as y*) for the group (A–H) despite it being more parsimonious to interpret y as three independent, parallel gains.

the unexpressed capacity to develop y. The underlying synapomorphy thus remains 'hidden', to be expressed within various subgroups only when it is selectively advantageous to the taxon.

Saether (1986) argued that Hennig (1966) recommended using underlying synapomorphies (which the latter termed homoiologies) for reconstructing phylogenetic relationships. However, Hennig treated homoiology as equivalent to convergence for this purpose and rejected it as a valid means of estimating cladistic relationships. While it may be true that the recurrent derivations of some character states may be caused by apomorphic tendencies of ancestral species, these tendencies cannot be used to provide evidence for grouping. In particular, there is no way of directly observing a taxon so as to determine whether it had an ancestor whose apomorphic tendency had not been expressed (Farris 1986a). Thus, by allowing the use of unobserved features in unobserved ancestors, the method of underlying synapomorphy provides a licence to group in any way one pleases. In Fig. 3.14, it is perfectly permissible to use the independent occurrence of state y in taxa B, C, and H to support their individual status as monophyletic groups, although these three hypotheses have relatively low explanatory power. However, the observed distribution of state y does not provide any evidential support for the monophyly of the study taxon A–H as a whole.

4.

Tree-building techniques

Ian J. Kitching

4.1 PARSIMONY CRITERIA

Parsimony criteria as used in tree-building methods attempt to minimize a quantity known as the optimality criterion. The decision as to which optimality criterion is to be used depends upon which underlying model is considered to be most appropriate for the data being analysed. This chapter explores optimality options and how they have been implemented into cladistic algorithms.

4.1.1 Wagner parsimony

Wagner parsimony is based upon concepts developed by W. H. Wagner (for example Wagner 1961, 1963). It was first formalized by Kluge and Farris (1969) for rooted trees and then generalized to unrooted trees by Farris (1970). Wagner parsimony is one of the two simplest procedures and imposes minimal constraints upon permitted character state changes. States must be measured on an interval scale, and thus binary, multistate and continuous characters can be used. However, multistate characters must be additively coded. Free reversibility of characters is allowed. For binary characters, therefore, the probability of a change from state 0 to state 1 is equal to a change in the opposite direction from 1 to 0. In the case of multistate characters, the probability of transformation to the next state in a sequence is equal to the probability of reversal to the previous state. For example, the probability of change from state 2 to 3 is equal to that for change from state 2 to 1. A consequence of this reversibility is that the length of the tree is independent of the position of the root. An unrooted tree evaluated using Wagner parsimony can be rooted at any point without changing its length.

In order to determine the minimum number of changes for a character using Wagner parsimony, only a single pass through the tree is required, beginning with the terminal taxa and proceeding to the root. Such a pass is termed a post-order traversal and is illustrated in Fig. 4.1.

The unrooted tree is first rooted by arbitrarily choosing one of the terminal taxa to act as the root (Fig. 4.1b). In practice, an outgroup would usually be chosen to fulfil this role.

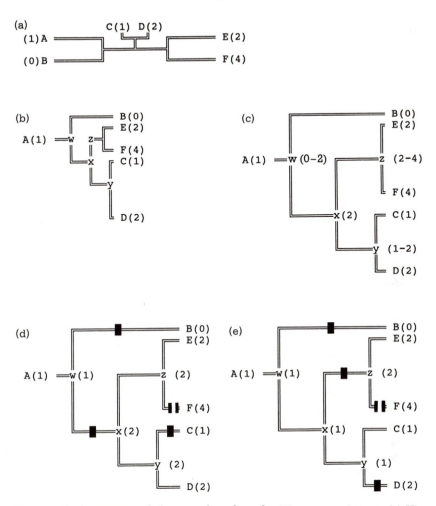

Fig. 4.1 Determination of character length under Wagner parsimony. (a) Un-rooted tree for six taxa. (b) Fig. 4.1a arbitrarily rooted using taxon A. (c) State set assignment at internal nodes by postorder traversal. (d) Nodal state assignments after preorder traversal. (e) Nodal assignments for an alternative MPR.

The optimization begins at the tips of the tree by choosing a pair of terminal taxa, E and F, which are linked by internal node Z. The state(s) (termed the state set) to be assigned to Z is calculated as the intersection of the state sets of E and F. In this example, the intersection of (2) and (4) is empty. Therefore, the smallest closed set that contains an element from each of the derivative state sets is assigned to Z, i.e. (2–4), and a value of 4–2 = 2 is added to the tree length. We then proceed to the next node towards the root but here only one of the derivative internal nodes has had its state set assigned it is necessary to pause and choose another pair of terminal taxa, in this case, C and D. The state set of their linking internal node, Y, is then calculated as the intersection of (1) and (2). In this case too, the intersection is empty and thus (1–2) is assigned to Y and 2–1 = 1 is added to the tree length. Proceeding to the next internal node, X, the state sets for both its derivative internal nodes have now been defined. The state set of X is then calculated as the intersection of those of Y (1–2) and Z (2–4). In this case, the intersection is not empty and therefore X is assigned to the value of this intersection, (2), and no increment is made to the tree length. Consider now the internal node W. For this node, again the intersection of the state sets for its derivative nodes, B (0) and X (2) is empty, and so W is assigned the state set (0–2), and 2–0 = 2 steps are added to the tree length. Finally, the state present in the root taxon A is examined to determine whether or not it is included within the state set assigned to its derivative node, W. In this example, it is included and therefore no further change is made to the tree length. Had the state in A not been included in the state set for W, then the tree length would have been incremented by the difference between that value and the nearest value in the state set of W. For example, had taxon A had state 3 instead of 0, then the tree length would have been incremented by 3–2 = 1.

In this way, the minimum number of steps required by a character on a tree is evaluated. In the example (Fig. 4.1c), five steps are required. However, this method does not unambiguously assign states to all of the internal nodes; that is, it does not produce an MPR (Swofford and Maddison 1987). In order to achieve this, a second pass through the tree must be performed in the opposite direction, i.e. a preorder traversal.

Passing from the root to the tips, each internal node is visited in turn. If a node has an ambiguous state set, then it is assigned the state of that set that is closest to the state found in its immediate ancestor. Applying this procedure to the tree in Fig. 4.1c, the states assigned to the internal nodes W, Y, and Z are 1, 2, and 2, respectively (Fig. 4.1d). Now, the evolution of the character can be traced on the tree. Maintaining A as the root, the character first changes by reversal state 0 between the internal node W and taxon B. It then also transforms to state 2 between node W and node X. The character remains unchanged between X and both Y and Z but then reverses to state 1 in C and transforms to state 4 in F. In the latter transformation, because Wagner parsimony requires characters to be additively coded, the change from state 2 to state 4 proceeds via state 3 and therefore involves two steps on this branch. Thus the five steps required by the character can be accounted for. The above procedure was described in set theory notation by Swofford and Olsen (1990).

It should be noted that this procedure (Farris 1970) will give a single, unique MPR only when all characters are free of homoplasy. In the presence of homoplasy, more than one MPR may exist, only some of which will be found by Farris' method. For example, the MPR shown in Fig. 4.1e, where there are independent transformations from state 1 to state 2 between nodes X and Z and node Y and taxon D, also requires five steps. An exact algorithm for obtaining all MPRs for discrete character data under the Wagner parsimony criterion was described by Swofford and Maddison (1987).

4.1.2 Fitch parsimony

For additively coded characters, where transformation from one state to another also implies transformation through all intervening states, the Wagner parsimony criterion is appropriate. However, a different criterion must be adopted for characters that are non-additively coded, in which transformation from one state to another is direct. Wagner parsimony was generalized by Fitch (1971) to produce a method that imposed no constraints upon the permissible character state changes. Fitch parsimony allows the free transformation of any state into any other state with the cost of only one additional step in tree length. Like Wagner parsimony, Fitch parsimony allows free reversibility of transformations.

Optimization using the Fitch parsimony criterion is similar to that described above for Wagner parsimony but with the following important differences.

1. Where the intersection of the two derivative state sets is empty, the state set assigned to the internal node is calculated as the union of the derivative state sets and the tree length is increased by 1.

2. If the state set of the root is not included in that of the next most distal internal node, then the tree length is increased by 1.

3. To obtain the MPR, if, when assigning unique states to the internal node, its state set includes the value assigned to its immediate ancestor, then this value is assigned to the node in question also. Otherwise, any value in the state set for the internal node is chosen arbitrarily.

The Fitch parsimony method, applied to the same tree as above, is illustrated in Fig. 4.2. The tree is again arbitrarily rooted using taxon A (Fig. 4.2b) and the state sets assigned to the internal nodes during the postorder traversal are Z (2,4), Y (1,2), X (2), and W (0,2) (Fig. 4.2c). The preorder traversal then assigns the unique states 2, 2, and 2 to the internal nodes W, Y, and Z respectively (Fig. 4.2d). Note that the assignment of state 2 to node W is purely arbitrary because the state found in taxon A ((1)) is not present in the state set for W ((0,2)). State 0 could equally well have been chosen for node W. Also, under the Fitch parsimony criterion, every state is only a single step from all other states. Therefore, the length of the branches linking node W to taxon B and node Z to taxon F are only 1 step long, as opposed to 2 under Wagner parsimony. Thus using Fitch parsimony, the character requires only 4 steps on the tree (Fig. 4.2d).

As for Wagner parsimony, this procedure for evaluating the MPR under Fitch parsimony need not necessarily produce a unique result.

An alternative MPR is shown in Fig. 4.2e, in which state 1 is assigned to node W. That this is a possible state assignment is not readily apparent from the original state set, (0,2), assigned to that node. To discover all possible MPRs, a second pass over the tree is necessary. One method and an algorithm for enumerating all possible MPRs was described by Fitch (1971). Because the states of characters under Fitch parsimony can transform to any other state with only unit increase in tree length, there are generally more alternative MPRs possible using this criterion than for Wagner parsimony.

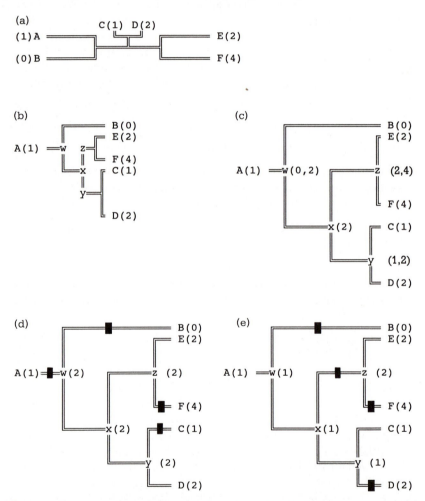

Fig. 4.2 Determination of character length under Fitch parsimony. (a) Unrooted tree for six taxa. (b) Fig. 4.2a arbitrarily rooted using taxon A. (c) State set assignments at internal nodes by postorder transversal. (d) Nodal state assignments after preorder transversal. (e) Nodal assignments for an alternative MPR.

Both the Wagner and Fitch parsimony algorithms described above are restricted to strictly bifurcating trees. However, they can be easily modified so that the MPRs of trees containing polytomies can be evaluated. Such modifications, using a variety of evolutionary models, were reviewed by Maddison (1989).

The above two parsimony criteria are appropriate where either the probabilities of character state change are unknown (which is usually the case) or where they are symmetrical; that is, where the probability of a change from 0 to 1 in a given time is the same as that for a change from 1 to 0. However, there are special circumstances, such as for certain types of molecular data, in which asymmetrical probabilities for change between different states may be more appropriate (see also Chapter 7).

4.1.3 Dollo parsimony

For data where the probability of a reverse change (1 to 0) is considerably higher than for a forward change (0 to 1), the Dollo parsimony criterion (Dollo 1893), as implemented by Farris (1977*b*), is appropriate. In its original formulation, Dollo parsimony differs from both Wagner and Fitch parsimony in requiring character polarity to be prespecified. It agrees with these two criteria in that the optimal tree is that which requires the fewest steps, but has the additional constraint that each derived character state is uniquely derived. Thus, all homoplasy must be accounted for by subsequent reversal to more plesiomorphic conditions. Multiple origins of the derived state(s), either by convergence or parallelism, are not permitted. DeBry and Slade (1985) considered that Dollo parsimony was an appropriate criterion under which to analyse restriction endonuclease cleavage map data for mitochondrial DNA. In such data, there is a large asymmetry between the probabilities of gaining a new site and losing an existing site, with the latter being very much more common.

To assess the number of steps required by a character under the Dollo parsimony criterion, only a postorder traversal of the tree is required. The procedure is illustrated in Fig. 4.3. The tree is first unrooted (Fig. 4.3a), then rerooted using one of the terminal taxa with the most derived state (Fig. 4.3b). This convention is adopted to make the subsequent procedure simpler. State sets are assigned to the internal nodes using the following rules.

1. If the state sets of the two derivative nodes are equal, then this value is assigned to the ancestral node and the tree length is not incremented.

2. If the state sets are different, then the higher value is assigned to the ancestral node and the tree length is increased by the difference between the two derived state sets.

3. When the basal internal node is reached, its state set is compared with that of the root. If they differ, the tree length is incremented by the difference. Otherwise no action is taken.

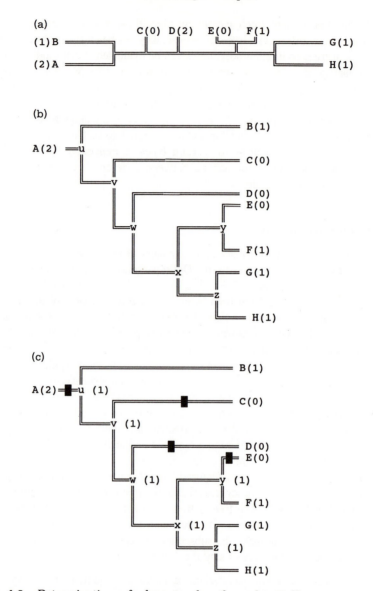

Fig. 4.3 Determination of character length under Dollo parsimony. (a) Unrooted tree for eight taxa. (b) Fig. 4.3a rooted with a taxon exhibiting the most derived state. (c) State set assignments at internal nodes.

Applying these rules to the tree in Fig. 4.3b, the character is found to require four steps (Fig. 4.3c).

The above procedure demonstrates that it is, in fact, unnecessary to specify the root in order to implement Dollo parsimony and a generalized unrooted

Dollo model has been developed (Swofford 1990; Swofford and Olsen 1990). Under the unrooted Dollo model, character states are assigned to internal nodes such that along a path connecting any two terminal taxa, a reverse change (from apomorphic to plesiomorphic state) is never followed by a forward change (plesiomorphic to apomorphic). In this way, the Dollo constraint, that each derived state should be uniquely derived, is not violated. An alternative formulation (Swofford 1990) states that the Dollo constraint is upheld if it is possible to construct a path between any two nodes that have the same character state such that the path does not pass through an intervening node that possesses a less derived state. Thus it can be seen that no matter where the tree in Fig. 4.3a is rooted, only four steps are required, with no more than a single transformation each to the apomorphic states 1 and 2 (for example Fig. 4.4a). It should be noted that this is true even when the tree is rooted between two nodes each of which show state 1 (Fig. 4.4b). In this case, however, state 1 is considered to be plesiomorphic for the group A−H. State 0 is then uniquely derived (above the branch bearing taxon G), followed by two reversals to state 1 and a subsequent unique origin of state 2. The tree in question therefore still only requires four steps, although the MPR is different.

If, however, a hypothetical ancestor showing state 0 is attached as the root, then the tree which was initially rooted between two taxa bearing state 1 (Fig. 4.4c) will be one step longer than that which was initially rooted using a taxon showing state 0 (Fig. 4.4b). The extra length in Fig. 4.4c comes from the initial gain of state 1 that has to be postulated between the root and the node subtending terminal taxon H. This example demonstrates that trees evaluated under the Dollo parsimony criterion are intrinsically rooted only if the state at the outgroup node (Maddison *et al.* 1984) is known or assumed (Swofford 1990).

The unrooted Dollo criterion is advantageous for restriction site analysis in that it dispenses with the need for a hypothetical ancestor. All that is required is the inclusion in the data set of one or more outgroups. Then if the derived state is absent within the outgroups, the analysis will postulate a single gain within the ingroup and then minimize the number of subsequent losses. If the derived state occurs also in the outgroup, then the most recent common ancestor of the ingroup will be postulated to have possessed the derived state and the analysis will seek to minimize the number of losses over the entire tree including the outgroups.

The disadvantage of Dollo parsimony is that if the assumption regarding the probability distribution being highly asymmetrical is false, it can severely overestimate the degree of homoplasy and thus the tree length (Swofford and Olsen 1990). Consider the tree in Fig. 4.5a. If the Dollo criterion is applied, then this tree requires seven steps: a single origin of state 1 followed by six reversals to state 0. However, if the probabilities have been incorrectly estimated and in fact are approximately equal, then Wagner or Fitch parsimony gives a tree (Fig. 4.5b) in which only two convergent developments of state 1 need be postulated. Under the latter probabilities, Dollo parsimony overestimates the tree length by

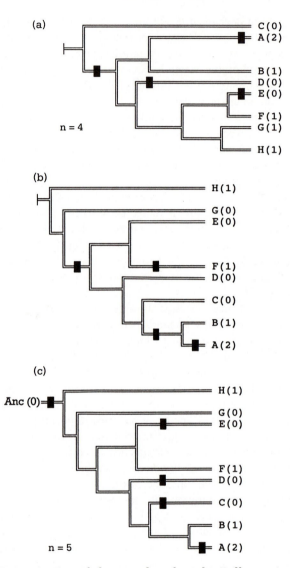

Fig. 4.4 Determination of character length under Dollo parsimony with generalized rooting. (a) Tree of Fig. 4.3a rooted on taxon C. (b) Tree of Fig. 4.3a rooted between taxa G and H. (c) Rooting the tree of Fig. 4.4a with a hypothetical ancestor (Anc (0)) requires an extra step.

five steps. The only means of avoiding this problem is to implement a 'relaxed' Dollo criterion, whereby one might prefer one gain and two losses to two independent gains but reject one gain and ten losses in favour of two independent gains. A method for implementing such assumptions is discussed under 'generalized parsimony' below.

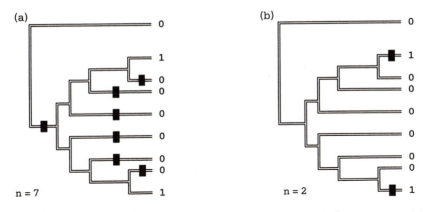

Fig. 4.5 Comparison of character length under Dollo and Fitch parsimony. (a) A tree requiring seven steps under Dollo parsimony. (b) The tree of Fig. 4.5a requires only two steps under Fitch parsimony.

4.1.4 Camin–Sokal parsimony

Another widely known parsimony method that imposes strong constraints upon the permissible character state changes, and was the first to be described for discrete character states, is Camin–Sokal parsimony. This method postulates that character evolution is irreversible (Camin and Sokal 1965). All homoplasy must therefore be accounted for by parallelism or convergence. Characters optimized under either the Camin–Sokal or Dollo criteria are examples of directed characters (Swofford 1990).

The method for assessing the number of steps required by a character under the Camin–Sokal parsimony criterion is very similar to that described above for Dollo parsimony. Again, only a postorder traversal of the tree is required. The procedure is illustrated in Fig. 4.6a,b, using the same topology and character state distribution as in Fig. 4.3a. Camin–Sokal parsimony only operates on rooted trees and the root must bear the putative plesiomorphic state (Fig. 4.6a). If the state of the root is not the plesiomorphic state, then the character polarity must be reinterpreted to make it so. State sets are assigned to the internal nodes using the following rules.

1. If the state sets of the two derivative nodes are equal, then this value is assigned to the ancestral node and the tree length is not incremented.

2. If the state sets are different, then the lower value is assigned to the ancestral node and the tree length is increased by the difference between the two derived state sets.

3. When the basal internal node is reached, its state set is compared with that of the root. If they differ, the tree length is incremented by the difference. Otherwise no action is taken.

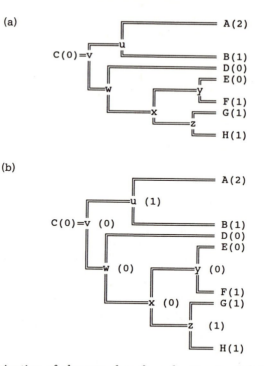

Fig. 4.6 Determination of character length under Camin–Sokal parsimony. (a) A tree of eight taxa; it must be rooted with a taxon showing the plesiomorphic state. (b) Nodal state assignments.

Applying these rules to the tree in Fig. 4.6a, the character is found to require four steps (Fig. 4.6b).

Unlike the other parsimony criteria described so far, Camin–Sokal parsimony is very rarely used. The assumption of irreversibility is very difficult to justify for any type of data, morphological or molecular.

4.1.5 Polymorphism parsimony

This parsimony criterion was developed independently by Farris (1978) and Felsenstein (1979). For binary characters, the method attempts to minimize the number of character state changes under the following constraints (Felsenstein 1989).

1. A polymorphism of states 0 and 1 is allowed to arise only once.

2. This polymorphism is maintained on the tree only as long as is necessary (i.e. the total extent of polymorphism required to explain the data is minimized), with homoplasy being explained by subsequent loss of one or other state.

Felsenstein (1989) implements this parsimony criterion only for rooted trees. However, polymorphism parsimony can also be applied to unrooted trees (hinted at by Felsenstein 1979, p. 60), using the following method.

The unrooted tree (Fig. 4.7a) is first rooted using a taxon that is coded 0 (Fig. 4.7b). This essentially arbitrary choice avoids subsequent conceptual difficulties regarding the usage of the terms 'maximum' and 'minimum'. Each terminal taxon and internal node is assigned two states, here referred to as L and R (for left and right). For each terminal taxon, the values of both L and R are equal to the state possessed by that taxon, while the root of the tree is assigned 0 for both L and R (Fig. 4.7b). A postorder traversal of the tree is then made. The values of L and R assigned to each internal node are, respectively, the maximum value of L and the minimum value of R possessed by its two derivative nodes (Fig. 4.7c). The subsequent preorder traversal leaves the values of R unaffected. However, the value assigned to L for a given internal node is set to zero if the sum of the values of L for its two descendent nodes and its immediately ancestral node is less than or equal to 1; otherwise it is left unaltered. Finally, if both L and R are identical, then that state is assigned decisively to the node; otherwise the node is polymorphic (Fig. 4.7d). Thus the tree shown in Fig. 4.7a requires four steps under the polymorphism parsimony criterion: the gain of state 1 to give the initial polymorphism, followed by two subsequent losses of state 0 and one loss of state 1 (Fig. 4.7e). An alternative interpretation, if states 0 and 1 represent alleles at a locus, is that an initial polymorphism evolves, followed by three subsequent fixation events. The procedure can be applied to multistate characters if these are first recoded into additive binary form.

This parsimony criterion should only be applied to characters that can show polymorphism within individuals, such as chromosome inversions (Farris 1978) or electrophoretic alleles.

4.1.6 Generalized parsimony

All of the above parsimony models, with the exception of polymorphism parsimony, can be considered to be special cases of a generalized method of parsimony. Under the generalized parsimony criterion (Swofford and Olsen 1990), a 'cost' is assigned to each transformation between states. A general algorithm for calculating the minimum number of steps and all MPRs of such a character was proposed by Sankoff (1975) and then developed into an exact, dynamic programming algorithm by Sankoff and Cedergren (1983). These authors described both the special case in which all the costs are equal to 1 (i.e. Fitch optimization) and then the general-case algorithm that is applicable to any cost values.

The costs are represented as a square m-by-m matrix, in which the elements, S_{ij}, represent the increase in tree length associated with the transformation from state *i* to state *j*, and m is the total number of states for the character. For Wagner parsimony (Table 4.1), it can be seen that the cost of transforming states through

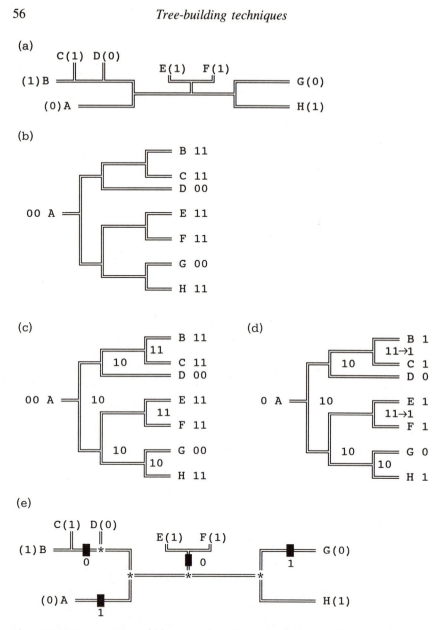

Fig. 4.7 Determination of character length under polymorphism parsimony. (a) Unrooted tree of eight taxa. (b) Fig. 4.7a rooted with a taxon showing 00. All terminals are labelled with their states. (c) Nodal state assignments after a postorder traversal of Fig. 4.7b. (d) Nodal state assignments after a preorder traversal. (e) Unrooted tree of Fig. 4.7a showing required changes. Nodes labelled with an asterisk are polymorphic (0,1). Labelled character state changes are fixations of the alternative allele distal to the label and gains to give a (0,1) polymorphism internal to the label.

the series is cumulative, whereas for Fitch parsimony, the cost of transforming between any two states is equal to 1. In the case of Dollo parsimony, M represents an arbitrarily large number that guarantees that each forward transformation occurs but once on the tree. The infinite cost of reversals in the Camin–Sokal matrix prevents such transformations from occurring.

The advantage of generalized parsimony is that it allows flexibility in permitted transformations that may not be available under any of the special parsimony criteria. For instance, several authors (for example Brown *et al.* 1982) have noted that in nucleotide sequence data, transversions (substitution of a pyrimidine for a purine or vice versa) occur much more rarely than do transitions (substitutions of a purine by a purine or a pyrimidine by a pyrimidine). Figure 4.8 illustrates a 'transversion parsimony' model in which transversions are allocated five times the cost of transitions. Under this model, it would require at least five transition substitutions to support an alternative topology before a topology supported by a single transversion substitution would be rejected.

Table 4.1 Comparison of m × m cost matrices for Wagner, Fitch, Dollo, and Camin–Sokal parsimony

	Wagner				Fitch				Dollo				Camin–Sokal			
	0	1	2	3	0	1	2	3	0	1	2	3	0	1	2	3
0	–	1	2	3	–	1	1	1	–	M	2M	3M	–	1	2	3
1	1	–	1	2	1	–	1	1	1	–	M	2M	∞	–	1	2
2	2	1	–	1	1	1	–	1	2	1	–	M	∞	∞	–	1
3	3	2	1	–	1	1	1	–	3	2	1	–	∞	∞	∞	–

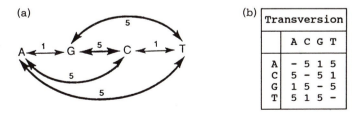

Fig. 4.8 Cost under transversion parsimony. (a) Model illustrating cost of transversion and transition substitution. (b) Cost matrix for Fig. 4.8a.

The cost matrix need not be symmetrical (i.e. the characters may be directed; Swofford 1990). Figure 4.9 illustrates a character in which forward transformations are optimized under the Wagner parsimony criterion, while reversals are optimized under the Fitch criterion. Generalized parsimony can also be used to implement a 'relaxed Dollo criterion'. The cost of a forward transformation is set to be greater than for a reversal but not so large as to prevent multiple gains altogether. For example, Table 4.2 illustrates a cost matrix where a single gain followed by multiple loss is the preferred hypothesis until the number of reversals exceeds four, when two independent gains becomes the preferred hypothesis. Several other models of character transformation are given by Swofford (1990).

There are, however, two problems in implementing generalized parsimony. The first is purely practical; the inclusion in a data set of characters coded using cost matrices greatly increases computation time. The second problem is that of determining the values that should be assigned to the elements of the cost matrix. This problem is analogous to the question of whether entire characters should be weighted a priori and is subject to the same criticisms (see also Chapter 5). Thus as a general rule, unless there are good reasons for applying differential costs (for example using transversion parsimony for nucleotide sequence data), then only Fitch and/or Wagner parsimony should be implemented.

(a)   (b)

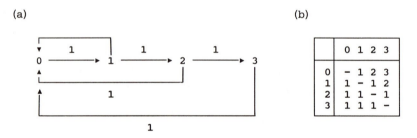

Fig. 4.9 (a) Model illustrating cost of transformation for directed characters. (b) Cost matrix for Fig. 4.9a.

Table 4.2 Cost matrix for 'relaxed' Dollo parsimony

	Relaxed Dollo	
	0	1
0	–	4
1	1	–

4.2 DELAYED AND ACCELERATED TRANSFORMATIONS

In an analysis of character evolution it is possible to utilize the relative amounts of parallelism and reversal as a means of choosing among competing MPRs. For example, while, a priori, one may not wish to prevent reversals completely, subsequently an MPR in which reversals are minimized may be preferred. By postponing character state changes as far as possible from the root of the tree, an MPR is obtained that maximizes the proportion of the homoplasy that is accounted for by parallelism. This is referred to as delayed transformation optimization (DELTRAN; Swofford and Maddison 1987; Swofford 1990).

If the converse requirement is chosen, that is, reversals are to be preferred over parallelisms, then the original optimization method of Farris (1970) can be used. Character state changes are then placed on the tree as close to the root as possible and homoplasy tends to be explained in terms of more distal reversals to plesiomorphic conditions. This procedure is also known as the accelerated transformation optimization (ACCTRAN; Swofford and Maddison 1987; Swofford 1990).

Use of these procedures to choose between competing MPRs is only justified when the root of the tree possesses the putative ancestral condition for each character. The evolutionary interpretation of the algorithms is lost if the root character states are chosen arbitrarily. Such would be the case if character polarities were not designated prior to the analysis (Swofford and Maddison 1987).

4.3 OPTIMIZATION OF 'MISSING' VALUES

A potential problem for optimization routines is that of 'missing data' (see also Chapter 8). Normally, every taxon in a data set will be allocated a state value for every character. However, there are two occasions when a specific numerical value cannot be assigned.

1. The structure is present in the study taxon but missing from the specimens examined because these are incomplete in some way. Perhaps the structure was broken off or was not collected; for example, characters of female morphology in taxa where the female is unknown.

2. The character may 'not apply', that is, the coding for the character is dependent upon those of other characters; for example, a character referring to conditions of wing venation must be coded as 'missing' for taxa that are wingless.

In either case, the solution is to assign to the taxon that state which is most parsimonious given the position of the taxon on the tree. Thus only those characters for which there are no missing values can affect the position of a taxon on the tree. In Fig. 4.10a, taxon B is placed by other characters as the sister group of taxon A and the missing value in B is therefore optimized to state 1. In the tree in Fig. 4.10b, a different character places B instead as the sister group of F. In this alternative, the missing value of B is optimized to state 0. In both cases, only 1 step is required for the character and a decision as to which of the two topologies is to be preferred rests with other characters.

There are often occasions when the assignment of a state to a taxon with a missing value is ambiguous. In Fig. 4.11a, B is optimized to state 1 and this state is thus interpreted as an apomorphy of taxa A, B, C, and D. But B can be equally parsimoniously optimized to state 0 (Fig. 4.11b), whence state 1 is interpreted as apomorphic for taxa A, C, and D only. An unambiguous assignment cannot be made under these circumstances.

From these examples, it can be seen that ambiguity in the form of missing values presents no problems for optimization procedures. While missing data are uninformative for actually placing a taxon on a tree, they may be useful in discriminating among tree topologies. Also, for values that are missing through being unavailable for coding, the optimized values can subsequently be tested when more complete data become available.

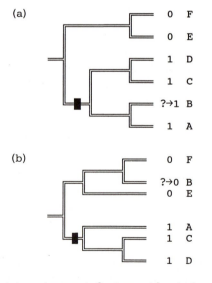

Fig. 4.10 Character state assignments for taxa with missing values {?}. (a) With taxon B treated as the sister group of taxon A, the missing value is optimized to 1. (b) With taxon B as the sister group of taxon F, the missing value is optimized to a 0.

4.4 X-CODING

Difficulties may occasionally arise in coding multistate characters for which there are two or more competing hypotheses of transformation available. For example, consider three taxa, A, B, and C, with states 0, 1, and 2, respectively. The first transformation series (Fig. 4.12a) proposes that state 2 evolved from state 0 via state 1. Using the Wagner parsimony criterion, the character states can be recoded into additive binary form (Fig. 4.12b). The cost of a change between any two adjacent states is thus 1, while the cost of changing between 0 to 2 is 2. An alternative hypothesis of character evolution postulates that states 1 and 2 have arisen independently from state 0 (Fig. 4.12c). When recoded in non-additive binary form (Fig. 4.12d), transformations between 0 and either 1 or 2 involve only one step, while changes between 1 and 2 involve 2 changes. (Note, therefore, that this form of non-additive coding does not give results identical to optimization under the Fitch parsimony criterion, which would postulate only one step to change from state 1 to state 2.)

If an analysis is performed that includes only one of these two hypotheses, the results will be biased towards that hypothesis. In order to assess the effects of the two hypotheses, and to discover which is more parsimonious, two analyses are required. However, as the number of such characters increases, the number

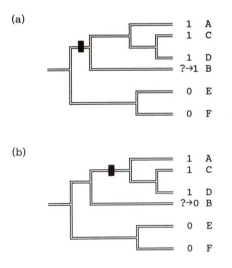

Fig. 4.11 Ambiguous state assignments for missing values. (a) Taxon B is assigned state 1. (b) Taxon B is assigned state 0. Both optimizations are equally parsimonious.

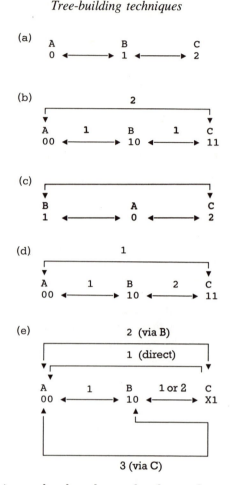

Fig. 4.12 Combining ordered and unordered transformation series using X-coding. (a) Ordered transformation series for 3 taxa. (b) Additive coding for Fig. 4.12a. (c) Unordered transformation series for 3 taxa. (d) Non-additive coding for a Fig. 4.12c. (e) X-coding resolution of the conflict between the transformation series of Fig. 4.12b and Fig 4.12d.

of analyses required to examine all the alternatives soon becomes prohibitively large. X-coding (Doyle and Donoghue 1986) is a method that allows the alternative hypotheses of character evolution to be combined, resulting in the need for only a single analysis.

To combine two transformation series using X-coding, the conflicting binary codings of one hypothesis are adjusted so as to make them consistent with the codings of the other hypothesis. This is achieved by replacing the conflicting codes with 'X'. Such coded data will then be treated by the computer algorithm as 'missing data' and global parsimony will determine which, if either, of the two conflicting hypotheses is to be preferred. In the example shown, the conflicting coding between the hypotheses in Figs 4.12b and 4.12d is the first binary code of C; which is 0 in Fig. 4.12d and 1 in Fig. 4.12b. Thus, this is the value that is replaced by X (Fig. 4.12e). If the resulting transformation series is examined, then it will be seen that the costs of changing between states are fully consistent with those postulated by both Figs 4.12b and 4.12d. For example, the path from A to C via B can be two steps, as in Fig. 4.12b, or three steps, as in Fig. 4.12d, depending upon the value that is assigned to the 'X' by the optimization procedure. The direct path from A to C is one step, as required by the hypothesis in Fig. 4.12d. The path from A to B via C is three steps. Doyle and Donoghue (1986) considered that this result introduced a subtle bias into their procedure because three steps are required whereas they considered only two need to have occurred. However, this conclusion is erroneous. If Figs 4.12b or 4.12d are examined, then it can be seen that three steps are required by both hypotheses in order to pass from A to B via C. This is entirely what is expected from a procedure that aims to render conflicting hypotheses of character state transformation consistent.

It should be noted that those parts of an X-coded transformation series that are recoded as 'X' play no part in the construction of the most parsimonious tree, merely assuming whatever value is consistent with that tree. Thus, as was noted by Doyle and Donoghue (1986), it is impossible to X-code all characters in order to avoid all possible bias, because doing so would remove all grouping information. They therefore recommended using X-coding only where there are major alternative hypotheses of character evolution.

4.5 SEARCHING FOR THE MOST PARSIMONIOUS TREES

Having selected a particular parsimony criterion, or combination thereof, as appropriate for the data to be analysed, the next problem is to find the optimal trees under this criterion. Methods for finding the maximally parsimonious or minimum length trees fall into two categories. For small to medium-sized data sets of up to about 20 taxa, exact methods can be used that guarantee the discovery of all optimal trees. For larger data sets, heuristic methods must be employed, which need not necessarily find all, or indeed any, of the optimal trees.

4.5.1 Exact algorithms

4.5.1.1 *Exhaustive search*
This method is the most simple to understand, in that all possible strictly bifurcating trees for a given data set are evaluated. A simple algorithm to perform this evaluation was outlined by Swofford and Olsen (1990) (Fig. 4.13).

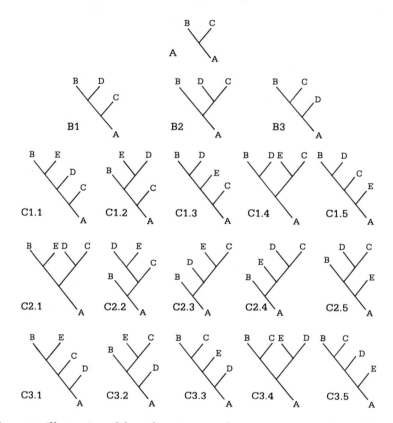

Fig. 4.13 Illustration of the exhaustive search strategy; see text for explanation (from Swofford and Olsen 1990).

Initially, the only possible tree for the first three taxa is constructed, which is termed the root tree (A). The fourth taxon is then added to give tree B1, followed by the fifth taxon to give tree C1.1. This process continues until all taxa have been added, giving a single tree. In this example, there are only five taxa, so the process terminates with tree C1.1. We then backtrack one node towards the root and add the last (in this case, the fifth) taxon in an alternative position, giving tree C1.2. This procedure continues until all possible derivative trees from tree B1 have been calculated (C1.1–1.5). We then backtrack one further node (in this example, to the root tree, A) and add the fourth taxon to a different branch to give tree B2. We then calculate all trees derivable from this new topology as before (C2.1–2.5). Finally, we backtrack once more to the root tree and calculate the derivative trees for the final path via B3 to trees C3.1–3.5. Calculation of the lengths of each of these fifteen trees will allow the identification of the optimal tree(s).

The difficulty with exhaustive search is that the number of trees increases rapidly with the addition of further taxa. For a tree currently comprising $i-1$ taxa, there are $(2(i-1))-3 = 2i-5$ possible positions to which the ith taxon could be connected. Thus, while there are 945 possible strictly bifurcating, unrooted trees for seven taxa, this value rises to over 2.2×10^{20} for a mere 20 taxa (Felsenstein 1978a). Thus, it is doubtful whether exhaustive search for data sets of more than 11 taxa is practical.

4.5.1.2 Branch-and-bound methods

However, there is an exact method available that does not necessarily require all possible trees to be evaluated — the branch-and-bound method — which was first applied to evolutionary trees by Hendy and Penny (1982). A simplified algorithm was described by Swofford and Olsen (1990). Initially, a tree is calculated, usually by one of the heuristic methods described below, the length of which is taken as the upper bound for trees subsequently generated by the branch-and-bound process. The sequence of tree building and evaluation then proceeds as for exhaustive search but the length of the tree is calculated as each new taxon is added. As soon as a tree is encountered where the length exceeds the upper bound, that path is abandoned because the addition of more taxa can only further increase the length. We can then backtrack one node and try a different path. In this way, the number of trees that must actually be evaluated can be significantly reduced.

If all taxa are added and the length of the tree equals the upper bound, then this tree is a member of the set of optimal trees. If, however, the length is less than the upper bound, then this tree represents the best found so far. The length of this tree is then substituted for the original upper bound and the process continued. This improvement is important because it may enable subsequent paths to be abandoned more quickly. When all possible paths have been evaluated, the minimum length trees will have been found.

The branch-and-bound method can be applied to data sets of up to about 25 taxa, although the exact number is highly dependent upon the efficiency of the algorithm used, the amount of homoplasy in the data, computer speed and patience of the analyst. There are several refinements used in the actual implementation to ensure early abandonment of path searches, including:

(1) using efficient heuristic methods to ensure that the initial upper bound is as close to the minimum length as possible;

(2) designing path search strategies so that highly divergent taxa are added early, giving fast initial increases in length; and

(3) using pairwise incompatibility to improve the lower bound on the length that will ultimately be required by trees descending from a tree at a given node (Swofford and Olsen 1990). These methods are described in more detail by Hendy and Penny (1982) and Swofford (1990).

In the worst-case implementation, in which most of a tree's length is concentrated in changes in the terminal branches, the branch-and-bound procedure is equivalent to an exhaustive search. Why, therefore, should an exhaustive search ever be performed? Simply, because the latter method provides access to suboptimal trees. Knowledge of these may be important for special applications, such as discovering how many trees are only one step longer than minimum, or calculating the frequency distribution of tree length, in order to discover where in that distribution a tree of particular interest lies.

4.5.2 Heuristic methods

The evaluation of most parsimonious trees is now known to be a member of a class of mathematical problems termed 'NP-Complete' (Graham and Foulds 1982). Consequently, it is unlikely that a general algorithm that can guarantee to solve all instances of the problem, such that the computational time is bounded by a polynomial function of the size of the problem, will ever be found. Such algorithms are termed 'effective' (Edmonds 1965). Therefore, in the absence of an effective algorithm, the time required to implement exact algorithms for NP-Complete problems increases extremely rapidly as the complexity of the data set increases. Very quickly, the time required to evaluate all the optimal trees for data sets of about 25 taxa or more (depending upon the level of homoplasy) becomes impractical. For such data sets, heuristic methods must be employed. These procedures search for the most parsimonious trees by approximate, trial-and-error techniques. Thus, they are not guaranteed to find the optimal trees.

Heuristic methods generally proceed by what is termed 'hill climbing'. An initial tree is taken and then rearranged in an attempt to decrease its length. When no further improvements can be made, then the process is stopped. However, we can never be sure whether the result thus obtained is the global minimum or merely a local minimum. However, the probability of locating the global rather than a local optimum can be improved by the application of two basic strategies: those that attempt to minimize the length of the initial tree by judicious addition of taxa during the building phase; and those that then rearrange that tree in order to try to decrease its length.

4.5.2.1 Addition sequences

Stepwise addition is the process whereby taxa are added to a developing tree. Initially, a tree of three taxa is chosen. A fourth taxon is then chosen and added to one of the three branches of the initial tree. A fifth taxon is then added to one of the five branches of this tree, and so on until all taxa have been added. However, there must be some means of determining which three taxa form the initial tree, and then which taxa are added in what order and to which branches. Several methods are available for determining the addition sequence of the taxa.

1. As is. The first three taxa are chosen to form the initial tree and the remaining taxa are then added in the order they appear in the data set. Unless the

optimal cladogram is nearly pectinate and the taxa appear on it in almost the same order as in the data set, this method is not very effective.

2. Random. In this method, a pseudorandom number generator is employed to reorder the taxa, which are then added to the tree using the as is method.

3. Simple. This is the method proposed by Farris (1970). As for the previous method, the addition sequence is determined prior to the commencement of tree building. First, a reference taxon is chosen; Farris (1970) referred to this taxon as the 'hypothetical ancestor' but any taxon in the data set can act as this reference point. In practice, an outgroup is usually chosen. Then, the Manhattan distance between this taxon and each of the remaining taxa is calculated (which Farris (1970) termed the 'advancement index'). The initial tree is then constructed for the reference taxon and the two taxa that are closest to it; i.e. have the two lowest advancement indices. Subsequent taxa are then added to the developing tree in order of increasing advancement index. Ties are broken arbitrarily.

4. Closest. This method differs from the previous three in that the order of addition of taxa to the developing tree is not predetermined before tree construction commences. First, the lengths of all possible trees of three taxa are evaluated and that with the shortest length is chosen as the initial tree. At each subsequent step, the increase in length that would occur by the attachment of all currently unplaced taxa to each branch on the developing tree is calculated. The taxon/branch combination that involves the least increase to tree length is then chosen as the next step. As for the simple algorithm, ties are broken arbitrarily. The closest procedure can thus be seen to require much more computation than do the other algorithms. In the latter methods, the number of tree lengths that must be calculated at any particular increment in tree construction is equal only to the number of possible attachment points (branches), because the taxon to be added is predetermined. The closest algorithm multiplies this figure by the number of unplaced taxa.

None of these methods works best for all data sets and there seems to be no means of determining which of these approaches will be best for any given application. The as is method is quick but may result in a tree that is quite far from optimal. Closest may result in a tree that is nearer to the most parsimonious tree but if the data set is large or complex, it may take an excessive amount of time to construct the tree because of the larger number of calculations that must be performed.

There is also the problem of 'islands of trees'. An island is defined as a group of tree topologies such that each member is no more than a single rearrangement away from another member of the set (Swofford 1990). ('Rearrangement' is defined under 'Branch swapping' below.) Generally, once one tree on the island has been found, then recursive branch swapping will locate all the other members. But it will be impossible, by definition, to reach a tree topology of another island by this process. Therefore, if the trees for a particular data set comprise a number of islands, and the topologies of these islands are very different, then unless the initial tree is a member of the island that also includes the optimal trees, those optimal trees will not be found.

However, the probability of locating the different islands can be increased by repeated branch swapping from different initial trees. Using randomly generated trees for this purpose would be inefficient, because such trees would very often be far from optimal, which would lead to protracted computation times. However, by using the random addition sequence, different starting trees could be generated for subsequent branch swapping, which while not being extremely far from the optimal trees, would be sufficiently different to reach alternative islands. Using the random addition sequence can also be used as a non-rigorous means of checking the efficiency of heuristic methods. Should 50 replicants give the same set of tree topologies, then it is likely that the maximally parsimonious trees have been found. However, if even after 100 replications, new topologies and islands are being discovered, then it is probable that yet further trees remain to be found.

A disadvantage of all the methods described above is that they will only retain a single tree at each step of the addition sequence, even if more than a single equally parsimonious partial tree is found. These alternative trees are the result of ties in the addition sequences simple and closest. Either there is more than one taxon that can be added at a given point, or there is more than one branch to which a single taxon can be attached, or both. Also, it is possible that although the addition of a particular taxon may be most parsimonious with regard to the partial tree, it may become a suboptimal choice when further taxa have been added. The end result then may well be a local rather than the global optimum. This is because all the methods aim for optimality for the current position; they cannot 'look into the future' to see what the results of a given action will be.

One means of circumventing this problem is to retain more than a single tree at each point during tree construction. These trees may be only the equally parsimonious or simply a fixed number, which may also include suboptimum topologies. Then by following a series of alternative tree construction sequences, the probability of finding the island that includes the optimum trees is increased. This procedure also decreases the effect of ties, because, to a certain extent, the alternatives are each followed through.

4.5.2.2 Branch swapping

However, unless the data set is nearly free of homoplasy, varying only the addition sequence is quite likely to result only in a local optimum. This unsatisfactory situation can be improved upon by the use of tree rearrangement algorithms, commonly referred to as 'branch swapping'. They are very much 'hit and miss' procedures but the hope is that if a shorter tree than the input tree exists, then these rearrangements will find it. Several methods of branch swapping, using different rearrangement criteria, are implemented in the computer packages available. Swofford (1990) briefly described the three that are used by PAUP v.3.0.

1. Nearest neighbour interchange (NNI). In this method, each internal branch of the tree can be considered to link two subtrees on one side to two on the other. In Fig. 4.14, the highlighted branch connects the two left hand side subtrees, (A + B) and C, to the two right hand side subtrees, D and (E +(F + G)). A nearest neighbour interchange exchanges one subtree on the left hand side with one on the right. In this case either D or (E +(F + G)) can be exchanged with C or (A + B) to give two possible NNI rearrangements (Fig. 4.14).

2. 'Subtree pruning and regrafting' (SPR). In this procedure, the tree is divided into two at a node; resulting in one subtree with a 'free branch' and one without. In Fig. 4.15, the subtree (A + B) is pruned and the branch connecting C with D on the other subtree straightened out. The pruned subtree is then re-attached by joining its free branch to either a terminal or an internal branch of the other subtree. In Fig. 4.15, the (A + B) subtree is reattached to the branch bearing taxon G. All possible combinations of subtree pruning and reattachment are evaluated.

3. 'Tree bisection and reconnection' (TBR). The process divides the tree into two subtrees by bisecting a branch between nodes. Both resulting 'free branches' are then themselves pruned, leaving two disjoint subtrees. Figure 4.16 illustrates this process, where the bisection occurs on the branch linking ((A + B) +C) to ((D + (E + (F + G))). The two subtrees are then reconnected by choosing one branch on each and creating a linking branch between them. In Fig. 4.16, the new branch is created between the branches bearing taxa B and G. All possible bisections and reconnections are evaluated.

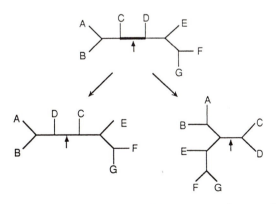

Fig. 4.14 Branch swapping: nearest neighbour interchange (from Swofford and Olsen 1990).

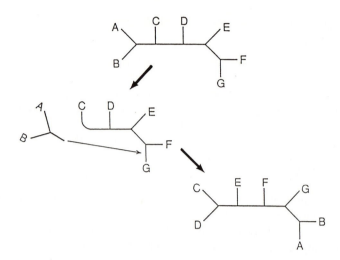

Fig. 4.15 Branch swapping: subtree pruning and regrafting (from Swofford and Olsen 1990).

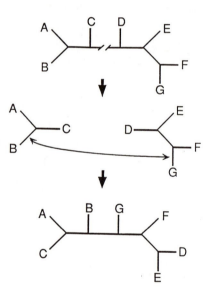

Fig. 4.16 Branch swapping: tree bisection and reconnection (from Swofford and Olsen 1990).

As alluded to above, these methods can only locate the maximally parsimonious trees if there is an unbroken series of rearrangements between them and the initial tree. If the series is broken, then locating the optimal trees will necessitate passing through intermediate trees that are longer than those that have already been found, resulting in entrapment in a local optimum. This problem can only be circumvented if options are available to allow branch swapping on suboptimal trees to be performed. A related problem concerns 'plateaux' in the set of tree topologies. In this case, the optimal trees are several rearrangements removed from the current position and all these rearrangements correspond to trees of equal length. Unless it is possible to retain all these equally parsimonious trees, rather than discarding all but the first one obtained because the rest do not represent improvements in optimality, then it will be impossible to reach the most parsimonious trees.

5.

Tree statistics; trees and 'confidence'; consensus trees; alternatives to parsimony; character weighting; character conflict and its resolution

Darrell J. Siebert

5.1 INTRODUCTION: TREE STATISTICS

This section outlines measures that describe certain aspects of trees or aspects of the relationship between a tree and the data set from which it was generated.

5.1.1 Tree length

The importance of tree length, used here as by Camin and Sokal (1965), stems from its role as the quantity minimized in the search for the 'best' branching diagram. Trees of different topology will be of different length if the data are informative, but if the data lack information to resolve relationships among the taxa then all trees will be of the same length. In short, when choosing among all the possible trees that could summarize the relationships among taxa of interest, the most parsimonious tree is the one of 'minimal' length. It is the 'optimal' tree. Tree length can be thought of as the number of 'steps' required to account for the data on the tree or the number of character state changes, or transformations, required to explain the data on the tree. Tree length is, in part, a function of the number of characters in a data set, and of the number of states of the characters. It increases as more characters, character states, or both are added to the data.

The concept of tree length is easily understood on a character-by-character basis. As an example, consider limbs among bony vertebrates (fins are plesiomorphic for the group). The tree of a trout, a coelacanth, a cow and a rat needs just one step, or character state change, to account for the tetrapod character legs (Fig. 5.1a). On this tree, the length of the character limbs is 1. The addition of a snake to the list of taxa under consideration requires an additional character state change, the loss of limbs, and the character is now of length 2 (Fig. 5.1b).

Adding a bat and a bird to the taxa under consideration and thinking of forelegs as 'wings' lengthens the character not by 1 step, but by 2 since 'wings' have to be accounted for twice, once in birds and once in bats (Fig. 5.1c). This oversimplification of some forelimbs as 'wings' demonstrates the effect of homoplasy on the length of a character, in that four steps are required on the tree (Fig 5.1c) even though there are only three character states. The length of

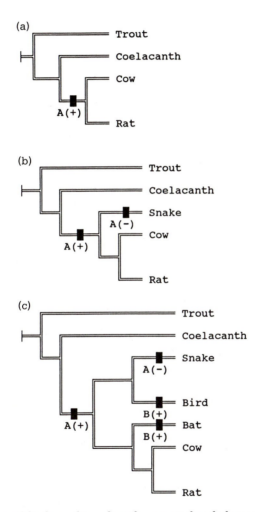

Fig. 5.1 Tetrapod limbs and tree length; accept the cladograms of vertebrates as given; paired fins are plesiomorphic for the group. (a) 'Legs' can be accounted for in one step. (b) Absence of limbs in snakes requires an additional step making the tree length 2. (c) 'Wings' in birds and bats require two additional steps on the tree because they are not closest relatives. Homoplasy lengthens the tree to four steps for three character states.

trees based on data sets containing more than one character is easily calculated as the sum of the length of the individual characters, provided the characters are independent.

Discussion of algorithms for calculation of tree length can be found in Farris (1970), Felsenstein (1978*a*), Swofford and Maddison (1987), Maddison (1989), and Swofford and Olsen (1990). In practice, calculation of tree length involves assignment of character state codes to terminals, then assignment of character state codes to internal nodes. The length of a branch (a segment between nodes) is calculated and added to the cumulative total of other branch lengths calculated previously.

5.1.2 Measures of fit between trees and data

The consistency index (CI), retention index (RI) and homoplasy excess ratio (HER) are measures designed to be descriptors of how well a tree describes a data set, or put another way, descriptors of how well the data fits a tree. Except for HER, these measures, like tree length, are summations of the measures for the individual characters over the suite of all characters in the data set.

The c.i. for a character is m/s, where m is the minimum amount of change possible for the character (m is equal to the number of states of the character minus one) and s is the actual number of changes in the character observed on the tree. Actual change, s, will exceed minimum possible change, m, to the extent extra steps, or homoplasy, are required to account for the character on the tree. Since s exceeds m in the event of homoplasy, c.i. has been used to measure homoplasy.

The ensemble consistency index CI for the tree is calculated by the summation of m (= M) and s (= S) over the suite of characters in the data set, such that CI = M/S. For a given data set CI = 1 when there is no homoplasy and decreases as homoplasy increases. Therefore, it has been used as an indicator of the level of homoplasy in the data; data sets with a high CI have been considered superior to those with a lower value. CI was intended for use in comparison among data sets and trees and its upper bound of 1 facilitates such use. However, CI has been found to be negatively correlated with both number of terminal taxa and characters. This detracts from its usefulness in comparisons among trees with different numbers of terminal taxa and characters (but see Goloboff 1991). It is also sensitive to uninformative characters (autapomorphies and symplesiomorphies) in the data set; they inflate CI without providing support for grouping of taxa. This makes its use as an indication of evidential support of groups (non-terminal taxa) somewhat problematic for those data sets that contain uninformative characters.

In Hennig86, Farris (1988) implemented what he termed the retention index (RI) to consider the fit between a data set and tree from a different viewpoint. The purpose of RI is to express the amount of synapomorphy in a data set by examining the actual amount of homoplasy as a fraction of the maximum possible homoplasy (symplesiomorphies and autapomorphies admit no possibilities of homoplasy so they do not contribute to RI). The amount of synapomorphy is measured as the complement of the measure of homoplasy. In effect, the RI can be thought of as the proportion of similarities on a tree interpreted as synapomorphy (Farris 1989). Therefore RI might be considered a better measure of the evidential support of groups than CI. The RI is high when state changes occur predominantly on internal nodes and low when changes are concentrated on branches leading to terminal taxa. The RI possesses the advantageous property of not being sensitive to uninformative characters; autapomorphies or symplesiomorphies will not inflate it.

The homoplasy excess ratio (HER) was proposed by Archie (1989) to address the perceived deficiencies of the CI as a tool for comparison among trees. His approach was to randomize character information to obtain an estimate of the behaviour of characters in such data sets so that behaviour could be compared with the behaviour of characters in real data sets. Observed homoplasy (the number of steps above the theoretical minimum number required) was compared with the theoretical maximum amount of homoplasy (the number of steps from the randomized data above the amount of the theoretical minimum). HER, or more appropriately the homoplasy excess ratio maximum (HERM) approximation of HER, is not dissimilar to RI, and Farris (1991) has discussed this and the perception of deficiencies in CI and RI (see also Goloboff 1991).

5.2 TREES AND CONFIDENCE

Systematists are often asked how much 'confidence' can be ascribed to a particular branching diagram. The question is largely unanswerable because most people rarely distinguish clearly between the psychological and statistical meanings of 'confidence'. Branching diagrams either correctly depict historical relationships, or they do not.

Some advocates of a statistical approach to phylogeny reconstruction have pushed the notion that we ought to develop measures of how much faith we can put in our trees. Confidence limits can be calculated for maximum likelihood estimations of branch lengths since they are statistical by definition. The situation is much more complex for trees. The problem centres on the complexity of trees; trees are so complex that the form of the distribution surface describing them is as yet unknown. This has profound implications. Thus far, statisticians have been unable to calculate true confidence limits for tree topologies or devise tests for different tree topologies. Felsenstein (1985), a strong advocate of the statistical approach, developed an analogue to the bootstrap to approach the problem of confidence in a tree. Others have approached the problem by data randomization (permutation) techniques. Goloboff (1991) has considered the problem from a quite different perspective, from the point of view of whether or not the data are decisive, that is, whether or not they permit a choice among trees.

5.2.1 Bootstrap and jack-knife

The bootstrap and jack-knife are techniques whereby an unknown distribution is estimated by repeated sampling from a sample distribution. The techniques have widespread use. Felsenstein adapted the bootstrap to arrive at 'confidence limits' for trees. He recommends randomly sampling either the character rows or columns in a data set to build a bootstrap data set of the same size as the original data set, which is analysed to give a tree(s). This procedure is repeated at least 100 times. The percentage of occurrences of a particular component that appears among the trees of the sample data sets can be considered an index of support (this technique does not result in true confidence limits in a statistical sense). Felsenstein, after a number of simulations, commented that for a component to appear within a 95 per cent 'confidence limit' it would have to be supported by at least three characters. It should be noted that the three character figure is imposed by the random resampling process, three characters being the number needed to get at least one included in the simulated data set. Sanderson (1989) has discussed bootstrapping as it applies to trees and suggested additional tests beyond that proposed by Felsenstein.

The jack-knife is a different type of resampling program, in that it is resampling without replacement and data sets are simulated by systematically leaving characters, or taxa, out of the simulated data sets. However, the justification problems are the same as those of the bootstrap.

5.2.2 Randomization

Faith and Cranston (1990) have suggested testing trees by examining the patterns of covariation among characters in random (permuted) data sets, the basic idea being that covariation of characters in real data ought to be much greater than that found in randomly generated data. This is not all that different in spirit from the idea behind HER. The test is designed to examine whether or not a data set has a hierarchical structure beyond that which appears in data sets in which characters covary randomly. The result, discovery of hierarchical structure (or lack thereof) in a data set, will depend on the choice of the null model (permutation strategy) employed. This choice, then, should be considered carefully.

As originally proposed by Faith and Cranston (1990), data sets of identical size and identical character state structure are created with character states assigned to taxa randomly within characters. These data sets are then used to search for the shortest tree. If the original data set produces a tree shorter than, say, all but 5 of the 100 random data sets, it can be certain at the 95 per cent level that the original data set has more hierarchical structure than would be expected by chance. This particular application tests the hierarchical structure of an entire data set. Responding to criticisms that a subset of a data set might possess hierarchical structure even when the data set considered as a whole might not, Faith (1991) proposed conditional tests to test the monophyly of a particular taxon within the group of taxa under study.

5.2.3 Data decisiveness (DD)

DD is based on the observation that data sets with a lot of homoplasy (low CI or RI) are not necessarily uninformative (Goloboff 1991). Data are considered informative to the extent that some cladograms are more efficient summaries of them than are others. This is measured by cladogram length. Data are informative if one or more cladogram explaining them is shorter than the others. Data are strongly decisive if one or more cladogram explaining them is very much shorter than others, and only weakly decisive if all possible cladograms are not very different from each other in length. This approach to judging cladograms is very different from the permutation strategy discussed above. The permutation strategy is premised on identifying the covariation among characters beyond that which might accrue purely by chance, and that is why it has been employed to effect a 'significance test'. Agreement among characters may arise by chance and there is no way to separate such agreement from that arising from history. DD measures the ability of data to resolve relationships among taxa without regard to its possible origin by chance.

   DD = $(\bar{S}-S)/(\bar{S}-M)$, where S = length of the most parsimonious cladogram, $\bar{S}$ = mean length of all possible cladograms for the data set and M = minimum possible variation. DD varies between 1 and 0. It is high for decisive data (a value of 1 is achieved when the data set has no internal conflicts) and low for data that do not permit much choice among cladograms (a value of 0 is achieved for wholly indecisive data).

5.3 CONSENSUS TREES

Trees generated from the analysis of a data set are called fundamental; they summarize a data set. Consensus trees are of another class of branching diagrams, those which can be considered as derivative trees. Derivative trees are constructed from a set of trees rather than from a data set; they summarize a set of trees. Summarizing sets of trees has proven useful in systematics, and especially so in biogeography. One application of a consensus tree is to summarize as a single tree the multiple solutions sometimes realized from the analysis of a data set. A perhaps more useful application is the summary of sets of trees from different data sets, for example combining the results from separate analyses of data sets of early ontogenetic stages and data sets of adult morphology. Results from molecular studies might be combined with morphological studies in this way too (Hillis 1987). There are many ways to generate a consensus tree from a set of trees. Only strict, Nelson, Adams, and combinable components consensus trees will be demonstrated here.

5.3.1 Strict consensus

The most conservative method is that known as strict consensus. Consensus trees are obtained by combining comparable groups (components) from two or more trees. A strict consensus tree is derived by combining only those components from a set of trees that appear in all of the original trees (Fig. 5.2). The rationale for strict consensus is that it includes only those components that are totally unambiguous and about which the data are absolutely clear. The polytomies present do in fact allow for any resolution present in any of the original trees. The highly conservative nature of strict consensus sometimes means that one is left with little resolution. For example, in Fig. 5.2 component 6 is the

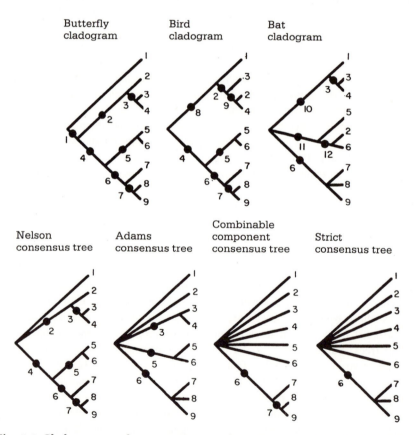

Fig. 5.2 Cladograms and consensus trees (from Bremer 1990). The butterfly, bird, and bat cladograms summarize different data sets. The four kinds of consensus trees summarize the different tree topologies of the butterflies, birds, and bats using different rules for their construction (see text for explanation).

only one common to all three trees and is thus the only one to appear in the strict consensus tree.

5.3.2 Combinable components consensus

A combinable components consensus tree is derived by including (or combining) those components from a set trees that are not contradicted (Fig. 5.2). Components need not therefore be included in all trees to appear in the consensus tree. Combinable components consensus may result in a tree with greater resolution than could be achieved with strict consensus, and sometimes more than any one of the original trees. If dealing with more than one data set, combinable components consensus can unite resolution unique to any of the data sets into one tree. Thus, in Fig. 5.2 component 7 is present only in the butterfly and bird cladograms. Resolution of this component in the combinable consensus tree does not conflict with the unresolved trichotomy of taxa 7, 8, and 9 in the bat cladogram, so it can therefore be admitted in the consensus tree.

5.3.3 Nelson consensus

There has been some controversy as to what constitutes a Nelson consensus tree. Some have mistaken it for strict consensus (see Page 1989*a*), and Nelson's (1979) presentation of consensus procedures seems to be a version of combinable components consensus. The currently accepted notion of Nelson consensus can be attributed to Page, who likened Nelson's discussion of the relations among components to clique analysis. Page then extended Nelson's analysis, which did not include contradictory replicated sets, to a formal clique analysis of cladogram components. Since the consensus tree is the maximum clique (or cliques if there are alternatives) some components can appear in the consensus tree that conflict with other components in the original trees. For example, in Fig. 5.2 component 3 occurs in both the butterfly and bat cladograms but conflicts with component 9 in the bird cladogram. Similarly, component 4 appears in the butterfly and bird cladograms but is not present in the bat cladogram due to the alternative position of the terminal taxa 5 and 6. Consequently, components 3 and 4 appear in the Nelson tree since they represent the largest cliques of components from the set of three trees under consideration.

5.3.4 Adams consensus

Adams consensus trees (Adams 1972) are designed to find the maximum number of components for a given set of cladograms by placing conflicting taxa at the most resolved node common to all trees (Fig. 5.2). For example, in Fig. 5.2 taxon 2 appears in three quite different positions in the three cladograms. Its position cannot be resolved so it is placed at the basal node in the Adams tree. Moving taxon 2 to the base recovers component 3 in the bird cladogram. As component 3 is now present in all three cladograms it appears in the Adams consensus tree. A similar argument can be applied to the recovery of component 5 in the bat cladogram, such that it also appears in the Adams consensus tree. Component 6 is present in all three trees so it too appears in the Adams consensus tree. One of the problems with Adams trees is that components which are not present in the original trees can appear in them. This is an undesirable property.

5.4 ALTERNATIVES TO PARSIMONY

Methods other than the parsimony method outlined earlier (see also Chapters 1 and 3) have been developed for the purpose of phylogeny reconstruction. Two methods, clique analysis or compatibility analysis and the three-taxon statements approach (TTS), claim to be parsimony methods. Likelihood techniques for phylogeny reconstruction also have been developed. Some of these alternatives to parsimony have been proposed because of perceptions of defects in the parsimony method. Such justification of these methods by reference to alleged defects in the parsimony method rather than as logical defences of systematics in their own right leaves them somewhat vulnerable to criticism. Farris (1983) showed that parsimony is to be preferred simply because it is the soundest method if we believe phylogeny reconstruction has an empirical base. Parsimonious hypotheses of relationships fit the original data more robustly than less than parsimonious solutions which require a greater number of *ad hoc* explanations to account for extra steps on trees.

5.4.1 Cliques

One method of applying the parsimony criterion to tree construction is to find the largest set of characters or character state trees in a data set which do not conflict among themselves. Such sets of characters are termed cliques and the method for finding them is sometimes called 'clique analysis'. Those suites of characters in a data set that do not conflict among themselves are considered 'compatible' with each other, so the method is sometimes called 'compatibility analysis'. Clique or compatibility analysis rests on the theory that characters arising through phylogeny cannot conflict, so they must be compatible with one another. Those that do must be 'mistakes' and thus can be ignored. Proponents of the method suggest that the largest set of non-conflicting characters are most likely to reflect phylogeny.

The approach originated in the pair-wise compatibility test of LeQuesne (1969) and algorithms for its implementation were described in Estabrook *et al.* (1976*a*,*b*). These authors proved the theorem that a group of two-state characters which were pair-wise compatible would be jointly compatible (Felsenstein 1989). Cliques can be calculated with the Fiala and Estabrook CLINCH program, a similar version of which is implemented in PHYLIP (Felsenstein 1989). More accessible paper-and-pencil applications were described by Meacham (1981), and Meacham and Estabrook (1985) and Felsenstein (1982) have discussed the methods in general terms and provided extensive lists of references. In the CLINCH algorithm binary characters are checked pair-wise for compatibility. After one pass through the matrix the program knows the largest clique (or cliques) of compatible characters which is then used to produce a tree. This may create a problem when multistate characters are recoded into binary form since part of the original character may be compatible whilst the rest is not.

Characters conflicting with the largest clique are not used in tree construction, a practice which has been criticized for throwing away or ignoring some of the data. In practice, the set of discarded characters may be large, even so large as to include most of the characters in a data set. Various attempts have been made at addressing this criticism by admitting characters back into the analysis after construction of a tree from the largest set of compatible characters (see for example Gauld and Underwood 1986; Sharkey 1989). The major criticism of this procedure is that the tree constructed from the largest clique may be quite unparsimonious globally because characters formerly excluded from the analysis can fit the tree only with extra steps.

5.4.2 Maximum likelihood

Felsenstein (for example, 1983) has been a long-term proponent of the view that phylogeny estimation should be viewed as a statistical problem, and tree construction should be treated as a problem of statistical inference. The major alternative is to see phylogenetic analysis from an hypothetico-deductivist point of view, most proponents of which approach phylogeny estimation by application of parsimony methods. This debate is far from settled and the practical problems of applying methods of statistical inference to tree construction may yet prove to be even more difficult than, say, parsimony methods. That maximum likelihood methods require a probabilistic model of evolutionary process prior to the analysis of pattern is viewed as problematic by proponents of parsimony methods. In truth, all statistical methods for phylogeny reconstruction require extensive knowledge of the evolutionary process, and it is precisely here that the weakness lies (Farris 1983). This knowledge of the evolutionary process is estimated in relation to a model of process.

Another current debate is whether or not we are engaged in classification or estimation of phylogeny (evolutionary history) directly. If it is the pursuit of the former, rather than the latter, cladogram construction with the aim of finding the most efficient data summary possible has advantages over all other methods proposed so far, and the arguments for statistical treatments lose force. Interpretation of cladograms as phylogenetic trees is possible once the most parsimonious cladogram has been obtained.

The maximum likelihood approach to phylogeny estimation is simply a method of discovering the tree that gives the highest probability of a data set being derived from it. It is not concerned with the probability of a tree being derived from a data set. The procedure requires one or more trees, a probabilistic model of evolutionary change and a data set. Given a tree and the model, the probability of the data set having resulted from that tree can be calculated. Note that the result is not the probability of the tree being correct (Felsenstein 1981*a*).

The result can only be as good as the accuracy and assumptions included in the model. The model has to describe the evolution of one character state from another across all taxa. For example, given DNA sequence data from a bacterium and a bird, the model has to be able to describe the conversion of one of the sequences into the other. To date, base substitution in nucleic acids, amino acid substitution in proteins and gene frequency changes are the only evolutionary changes simple enough to have been modelled, and recent (unpublished) tests demonstrate most of those for base substitution in nucleic acids do not give results within the expected range. At this point in time it appears that realistic probabilistic models of processes of morphological change are simply not possible, though they might be developed in the future.

Calculation of the probability of a data set having resulted from a given tree proceeds by summation of the probabilities calculated for each individual character over the suite of all characters in the data set. Given the large number of base sites in most nucleic acid data sets, the problem of obtaining this probability is not insignificant, even with the computers currently available, although the problem is tractable. One of the main advantages of maximum likelihood methods is that they are statistically consistent by definition, that is as the data sample tends to infinity, the result converges on the 'truth' of the model.

Applications of these techniques so far have involved data sets with as few as several hundred, to several thousand, base sites. The question of whether or not these data samples are large enough to have attained consistency is still open (it may be that sequencing entire genomes will not provide data sets of sufficient size to achieve consistency). The problem of tree construction from a data set by iteration with a maximum likelihood model is much more complex than obtaining a probability from a given tree. Felsenstein (1981*b*) has described a stepwise addition approach that is initialized with two taxa. The likelihood is calculated for these two taxa and a third taxon is added and the likelihood evaluated. Remaining taxa are added sequentially, trying all possible positions in the tree to maximize likelihood. Stepwise addition approaches are subject to error in that the best topology is not always found. The taxa used as the starting point can bias the outcome. Repeating the process a number of times with different starting taxa may help to find a topology with a greater likelihood. A real problem is that systematic problems often involve dozens of species. Finding the tree with the greatest likelihood from among the astronomical number of possible trees for large numbers of taxa remains unobtainable in practice. Maximum likelihood approaches are practical only for those problems involving a few taxa.

5.4.3 Three-taxon statements

Synapomorphies can be thought of as units of resolution in systematics. The smallest systematic problem in which a resolution is possible is for three taxa. Two taxa can only be related. Given three taxa, two of them might be each other's closest relative in comparison with the third. Any systematic problem of greater than three taxa, say five taxa, can be broken down into a series of three-taxon problems that can be added together to produce a resolution for the whole problem.

Recently, Nelson and Platnick (1991) have proposed a new approach to cladogram construction, one they consider potentially to be a better application of the parsimony criterion. This new method seems to include an element of cladogram information known as term information; present parsimony methods do not, they include only component information. Inclusion of term information seems to allow greater resolution in some instances. The three-taxon statement method is also thought to be a way to deal more accurately with missing data. It also seems a better way to deal with taxa for which some of the characters seem inappropriate, for example, when comparing invertebrates with vertebrates the vertebral column is an inappropriate character for the invertebrates.

The method involves breaking down groupings suggested by individual characters into allowed three-taxon statements, and coding these three-taxon statements in an expanded data matrix. The new matrix is processed with a standard parsimony algorithm to achieve one or more branching diagrams. This amounts to finding the largest set of allowed three-taxon statements (somewhat analogous to cliques). The result may be compared with the result obtained from a strict parsimony analysis of the original data by examining the number of three-taxon statements each will accommodate. The three-taxon statement method sometimes results in resolution that accommodates greater numbers of three-taxon statements, the implication being that it codes more accurately and accommodates more term information in characters than do other coding methods. It also sometimes provides more resolution than a strict parsimony analysis, by obtaining both fewer numbers of cladograms and greater resolution of taxon relationships.

This method seems interesting in its ability to incorporate term information, but the comparison of results obtained from it and from a strict parsimony analysis is not straightforward. The relationship between taxa and character distributions is clear in a strict parsimony analysis; it is not in the three-taxon statement analysis. A character indicating a relationship among a large group may be broken up because it is overpowered by the allowable three-taxon statements of mutually conflicting characters for groups of smaller size.

5.4.4 Phenetics

Phenetics, or numerical taxonomy, developed in the 1960s from the realization that phylogeny is mostly unknown and might be unknowable. If genealogy was unknowable, genealogical classification is impossible. Developing out of operationalism, numerical matrix techniques were originally developed in the late 1950s and early 1960s to allow classification to be based on overall similarity, the argument being that classifications incorporating overall similarity would be most informative. The principal aims of phenetics were to overcome intuitive methods and to be objective, explicit and repeatable, both in evaluation of taxonomic relationships and in the erection of taxa (Sokal and Sneath 1963; Sneath and Sokal 1973). This was achieved by producing consistent data matrices, examining large character sets, and weighting all data equally. Classifications were determined by producing similarity matrices from the original data matrices and dendrograms (branching diagrams) from the similarity matrices. Many kinds of similarity measures and clustering algorithms were developed, which eventually overshadowed the original aims of objectivity and repeatability. Even more critical is that phenetics has only a crude homology criterion of 1:1 correspondence and does not associate characters and taxa. Consequently, the ability to distinguish between homology and homoplasy does not exist, and hence similarity is untested evidence of relationship. It is precisely by showing that hypotheses about characters give rise to hypotheses about groups using the property of homology (similarity, conjunction, and congruence) and the parsimony criterion of choice between competing hypotheses of relationship that has allowed cladists to demonstrate the empirical basis of systematics. Farris (1977a, 1979b, 1980, 1982b) has shown more formally how cladistic classifications are more informative than phenetic ones.

   In spite of this, numerical taxonomy is still practised by some systematists, particularly in research of species complexes, within species variation, and groups such as bacteria, in which taxonomic research has been dominated by numerical phenetics. Centroids of clouds of points in morphometric multivariate space can only be associated by phenetic techniques, and the pair-wise distances resulting from immunological and nucleic acid hybridizations are phenetic. It should be remembered that these methods do not associate characters and taxa, and are thus susceptible to grouping by symplesiomorphy.

5.5 CHARACTER WEIGHTING

Character weighting is the process of assigning a 'valuation' factor to characters. Characters are thereby given relative ranks. Some characters will be 'valued' higher than others, and these will figure more prominently in the decision of which branching diagram is to be preferred.

Character weighting is a vexing subject. There are many approaches, most of which appeal to some external theory for justification with the effect that another layer of assumption is added to the analysis. For example, character compatibility has been suggested as a rationale for character weighting (LeQuesne 1969; Gauld and Underwood 1986; Sharkey 1989), as has the preference for use of likelihood methods for phylogeny reconstruction (Felsenstein 1981*a*). Additional assumptions further remove systematic analyses from their empirical basis, and affect the distribution of characters among taxa. In the author's view, this is good reason for not doing it all (but see Neff (1986) and Wheeler (1986) for an opposing viewpoint).

One aspect in the philosophical arguments over characters relates to their meaning. If one considers characters as synapomorphies, as has been suggested by Platnick (1979) and Patterson (1982*a*), then weighting is problematic because it implies that some synapomorphies are of greater value than others, an implication for which there appears to be no obvious justification. If a character is nothing more than any particular feature of an organism then some form of character weighting might be justified, since it is recognized that not all features of the organisms under study will contribute information to the resolution of the relationships among them. These uninformative, or less informative, features are then candidates for low, or lower, weights. No weighting protocol yet proposed to achieve this end has been acceptable to all. Two approaches will be discussed briefly below. The extreme case is claimed to be the position of the cladists where their refusal to admit data that is irrelevant (contains no information bearing on the problem) to the analysis is called zero weighting of characters.

Farris (1969) presented an interesting approach, which he called successive approximations weighting. It is a procedure based on a notion of 'cladistic reliability', or more plainly, the degree to which a character conforms to a hierarchy. With regard to avoiding unnecessary initial assumptions in a systematic analysis, this weighting protocol is related to performance of the characters after a round of analysis using equal weights. The degree to which a character conforms to a hierarchy is related to homoplasy. Farris used the consistency index of a character (c.i.) as the weighting factor since it is related to homoplasy of a character. A character with perfect fit to a tree has a c.i. of 1 and those which require homoplasy have a progressively lower c.i. in proportion to the number of extra steps required to make the character fit the tree.

Characters consistent with a hierarchy receive higher weight and those inconsistent with it receive lower weight. The protocol is iterative. A data matrix is analysed to find the most parsimonious tree, then the c.i. of each character is calculated, and then used to define the weight of each character in the next round of analysis. The weighted data matrix is now analysed for the most parsimonious tree and this new tree is compared with the first. If it differs, the c.i. is recalculated and applied as a weight for another round of tree search. The process stops when the form of the tree no longer differs between iterations. In practice, successive weighting usually reduces the number of equally parsimonious cladograms, but can sometimes result in a proliferation of cladograms.

In Hennig86 Farris (1988) refined the approach by using, the rescaled consistency index ($rc = ci \times ri$), as the value of weight since it achieves a value of 0 for completely homoplasious characters, which the ci cannot. However, Goloboff (1991) points out that even when the aim of eliminating a completely homoplasious character has been achieved with the rc, rather than with the ci, in successive weighting, the use of rc as a weighting function is still problematic because it does not always increase with less homoplasy. In an extreme case, successive weighting with rc results in weights of either 0 or 1 being assigned, which has been likened to the effects of clique analysis.

Sharkey (1989) proposed another weighting method using compatibility analysis. As described earlier, the maximal clique is first determined, but instead of rejecting the excluded characters, these are weighted in relation to their degree of incompatibility, taking into account the degree to which characters might be incompatible purely by chance. These are then added, one by one in reverse order to the cladogram derived from the maximal clique from which they were first excluded, in as parsimonious a fashion as possible without changing the topology of the maximal clique. More germane to this discussion, Sharkey also suggested applying the weights to parsimony analyses. The weights are applied to the data matrix, which is then analysed using a parsimony algorithm. However, it is worth noting that the weighting technique is character compatibility, which amounts to resorting to an external theory for justification. Use of a parsimony algorithm in a late stage of analysis does not render the same result as would be obtained from a global parsimony analysis.

5.6 CHARACTER CONFLICT AND ITS RESOLUTION

Character conflict, or homoplasy, occurs in almost all real data sets. These conflicts present dilemmas for which there are no easy solutions. Some conflict will resist all attempts at resolution, and *ad hoc* postulates of homoplasy are employed to explain the distribution of conflicting characters among taxa. Conflicting characters are also 'falsifiers' in the hypothetico-deductivist approach to phylogeny reconstruction; they are evidence that the chosen phylogeny might be incorrect. One particularly vexing manifestation of character conflict is that there may sometimes be several, or even hundreds of trees from one data set. There seem to be four main ways of dealing with character conflict: (1) re-evaluation of the characters, (2) character weighting, (3) transformation series analysis (TSA), and (4) consensus trees.

5.6.1 Re-evaluation of characters

Re-evaluation of characters falls conveniently into two approaches, that which is specimen based and those which are based on some sort of manipulation of the data matrix.

5.6.1.1 Specimen-based approaches

If the data are morphological characters then the first recourse should always be to check the original specimens. One potential explanation is observer error, and material rechecking may reveal that a mistake has been made. Similarity observed among the features of taxa under study may be found to be otherwise on closer examination. Also, study of the development of the features may reveal them to be of different origin. It may be possible to decide that the feature was inappropriate in the first place and that the states belong to a different character.

5.6.1.2 Data matrix manipulation approaches

Another approach to reconciliation of character conflict is through manipulation of the data matrix. Character weighting and transformation series analysis (TSA) are two such approaches.

Character weighting Character weighting was presented earlier as an application to set relative values to characters. Characters of high weight figure more prominently in the final choice of a tree, and this may result in one or fewer solutions in relation to the unweighted analysis. It must be remembered, however, that even if character weighting results in fewer trees, or even a single tree, the original observation of 'sameness' in seemingly unrelated taxa remains.

Carpenter (1988) suggested the successive approximations approach to character weighting (Farris 1969) as a possible solution to the problem of choosing among equally parsimonious trees. Recall that successive weighting is designed to devalue homoplasious characters, those that conform less to the hierarchy. The procedure is available in both Hennig86 (see above) and PAUP. Data are weighted by their rescaled consistency index (rc, see above) after the initial search for parsimonious solutions, and then re-searched. The procedure repeats until weights stabilize between searches. If all goes well, a single tree emerges, or at least fewer trees emerge than emerged in the initial analysis. In the cases in which the iteration process results in all characters being assigned a weight of either 1 or 0, the protocol devolves to a sort of clique analysis.

One disconcerting result is that the tree resulting from this procedure some-times is not among that set of trees derived from the initial analysis. Just how this new tree, based on weighted data, can be compared with the original trees is not clear, since the new tree is based on a different data set. The data set is different because the effect of weighting is to expand it, for example weighting a character by 3 has the effect of adding two additional characters to the data set, albeit with identical character state distributions among taxa as the original character.

An even more disconcerting result is that successive approximations weighting of some data sets with low initial CI can produce more trees than the initial, unweighted search. The conclusion here is that the data set probably does not contain information pertinent to the question. Calculation of the data decisiveness (DD) statistic would be appropriate to find out whether or not the data are decisive, even if homoplasious.

Transformation series analysis   Transformation series analysis (TSA) is an approach currently under development that potentially can be used as a means of resolving character conflict (Mickevich and Weller 1990; Mickevich 1982; Lipscomb 1990; Mickevich and Lipscomb 1991). Its use is restricted to the analysis of multistate characters. In particular, TSA will address the conflict resulting from ordering multistate characters. It attempts to bring the ordering of multistate characters into conformity with the hierarchy inherent in the rest of the data. TSA begins with the construction of an initial set of character state trees from which one or more trees are generated. Those character state trees that conflict with the most parsimonious tree are recoded to conform to adjacent positions on the obtained tree. The data set is then recoded and re-analysed to obtain a new tree. The process of comparing the character state trees, recoding, and new tree construction continues until a set of stable character state trees is achieved.

5.6.2 Consensus trees

Consensus trees can be considered as an indirect method of resolving character conflict in general classifications, since they reduce the number of trees under consideration to 1. The basic idea is to construct a tree from the non-contradictory components found among the set of trees generated from the initial analysis, the resulting tree being thought of as a compromise or 'consensus' solution. Recall that a consensus tree is a derivative tree constructed from a set of trees; it summarizes a set of trees, and as such may contain components not found in any of the original trees. Miyamoto (1985) and Carpenter (1988) have advised against their use for this reason, and because a consensus tree rarely summarizes a data set as efficiently as any of the set of trees from which it was constructed. The first criticism does not generally apply to the resolution of character conflicts, since it is made of those sets of trees which were generated from different data sets. However, the second criticism is one that should be considered (but see Anderberg and Tehler (1990) and Bremer (1990)).

6.
DNA analysis: theory

David M. Williams

6.1 INTRODUCTION

The use of molecular data in systematics has dramatically increased over the last two decades. Naturally enough, techniques for the analysis of the data have developed more slowly (Moritz and Hillis 1990). Although there are a wide variety of computer programs available for implementation, the underlying theoretical justification behind their use is often not fully discussed; programs are sometimes treated as 'black boxes' which produce *the* phylogenetic tree. There are a number of good discussions concerning the range of available programs, their algorithms, justification and use in the literature (Nei 1987; Li and Graur 1991; Swofford and Olsen 1990). This chapter discusses some generalities pertinent to molecular data, outlines the principles that have been employed in tree reconstruction methods, and finally describes some relevant programs. The reader is referred to the above references for further information.

6.2 HOMOLOGY

In Chapter 2 the concept of homology was discussed within the framework of morphological attributes of particular organisms. Some have argued that it may not be useful to use this classical anatomical term for contemporary molecular data (Aboitiz 1988). However, others have suggested that a more fruitful investigation may be the search for areas in which there is common ground (Patterson 1988*a*). This is the line followed in this chapter. In comparative biology homology is the central concept, it is at the heart of systematics (see also Chapter 1). The most common meaning of homology is common ancestry. However, this is an explanation of homology and does not necessarily guide us to its discovery (Patterson 1982*a*). Alternatively, homology can be viewed as a relation that specifies groups (taxa). In molecular systematics there are different levels at which homology can be established. For instance, there is the relation among genes from individual genomes, as in Fig. 6.1. Further, there is the more detailed correspondence between the amino acid and/or the nucleotide sequences within individual genes; when sequences from genes are written out they form

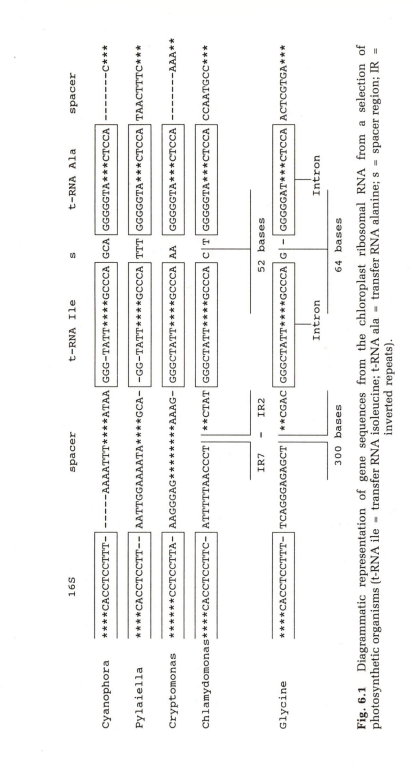

Fig. 6.1 Diagrammatic representation of gene sequences from a selection of photosynthetic organisms (t-RNA ile = transfer RNA isoleucine; t-RNA ala = transfer RNA alanine; s = spacer region; IR = inverted repeats).

a further series of potentially informative characters. Thus, there are two levels at which comparison is possible. This illustrates one difference between morphological and molecular homologies: in molecular systematics the genome is the 'unit' of comparison, where the genome is defined as the total array of individual genes, spacer and initiator regions (Patterson 1988a). At either level of investigation there is a series of tests that can help elucidate homology, distinguishing it from other relations (Patterson 1982a).

6.3 HOMOLOGY TESTING

6.3.1 Similarity

In the early molecular literature the term homology was used as a direct equivalent of similarity, as it sometimes still is in morphological systematics (Bock 1989). Supposing that the nucleotide sequences from the same gene in two different species exactly match, base for base, they would be considered 100 per cent homologous and as differences increase so the percentage decreases. Yet, this measure is based upon site identity; and similarity at site positions does not necessarily equate with homology, since homoplasy (false similarity), as conflicting evidence, almost always occurs (Fitch 1966, 1970). Thus, there is a degree of quality in similarities that divides them into informative (homologous) and uninformative (homoplasious) characters (Li and Graur 1991). However, unlike morphological characters, there can be no detailed study of similarity; bases are either identical or they are not. In morphology, one can examine the ontogeny of features for more detailed comparisons. A nucleotide base cannot be submitted to further investigation, whereas in morphology wings, for instance, may be determined to be 'different' when more detailed anatomical examination is made (see also Chapter 1).

6.3.2 Conjunction

The conjunction test is used in morphology to distinguish between two structures deemed to be homologous by first passing the similarity test. If both structures are present in the same organism at the same time then they cannot be homologues; this is exemplified in the case of serial homology where the relation is known as homonomy (Patterson 1982a). It would appear that because genes invariably occur in multiple copies all molecular relations would fail this test. However, using the haploid genome, rather than the entire organism, prevents this difficulty (Patterson 1988a). However, some genes do co-occur on the same genome and as a consequence fail the conjunction test. The results from such observations highlight a distinction between paralogous genes, which reveal the history of genes, and orthologous genes, which reveal the history of species (Fitch 1970); similarity alone is insufficient to distinguish these relations.

6.3.3 Congruence

Congruence is the most decisive test in morphology in that it separates those characters which are useful in systematics from those which are not (Patterson 1982a). Shared similarities are homologies that diagnose mono-phyletic groups and are consequently congruent with other homologies (see also Chapter 1).

6.3.4 Results of testing for homology

Applying Patterson's tests (above) to morphological comparisons (Table 6.1), the relations that pass the congruence test are those that are useful in systematics: homology, homonomy, the complement relation, and two homologies. Patterson (1988a) described five different relations in molecular systematics, some which have direct equivalents in morphology (Table 6.1; after Patterson, 1982a, 1988a) and others which are unique. These are described below and illustrated in Fig. 6.2a–c. With molecular comparisons, although four relations pass the congruence test (Table 6.2; orthology, paralogy, the complement relation, and two orthologies), one of these relations, paralogy, discovers gene phylogenies rather than species phylogenies (Table 6.2). Of the five relations which pass the similarity test (orthology, paralogy, xenology, paraxenology, and plerology) only one, orthology, contributes to species phylogenies, three (paralogy, xenology, paraxenology) contribute to gene phylogenies, and the remaining one, plerology, is problematic. Thus, molecular homologies can reflect the phylogeny of either species or genes.

Table 6.1 Comparison of molecular and morphological relations and their performance on Patterson's tests for homology; C = congruence, S = similarity, Cj = conjunction; (+) = pass; and (–) = fail. (After Patterson 1982a, 1988a)

Molecular	C	S	Cj	Morphological
Orthology	+	+	+	Homology
Paralogy	+	+	–	Homonomy
Complement	+	–	+	Complement
Two orthologies	+	–	–	Two homologies
Xenology	–	+	+	Parallelism
Paraxenology	–	+	–	Multiparallelism
Plerology	–	+	–	Homeosis
Convergence	–	–	+	Convergence
?	–	–	–	Endoparasitism

6.3.5 Orthology

Orthologous sequences (ortho = exact; Fig. 6.2a; left-hand tree) are homologues reflecting the descent of species (Fitch 1970). It is equivalent to homology in morphology and is the most useful relation for discovering species relationships.

6.3.6 Paralogy

Paralogous sequences (para = parallel; Fig. 6.2a; right-hand tree) are homologues reflecting gene history (Fitch 1970). They can co-exist in the same organism, such as the haemoglobin family (myoglobin, alpha, beta, gamma, delta, and epsilon chains; see Goodman *et al.* (1987)), and the phycobiliproteins of the chloroplasts of blue-green algae (Glazer 1987). Paralogy is explained by gene duplication and its equivalent in morphology is homonomy.

6.3.7 Xenology

Xenologous sequences (xeno = foreign; Fig. 6.2b) will reflect, in part only, gene history. However, the sequence will be incongruent with the organisms carrying the gene. Horizontal gene transfer or transfection are the assumed causes (Gray and Fitch 1983).

6.3.8 Paraxenology

Paraxenology ('duplicate or multiple xenology'; Patterson, 1988*a*) differs from xenology by the presence of two or more copies of the foreign gene in the host

Table 6.2 Molecular relations and their ability to resolve species or gene phylogeny, and their performances against Patterson's tests for homology. (After Patterson 1988*a*)

Molecular	Congruence	Similarity	Conjunction
Species phylogeny			
Orthology	+	+	+
Complement	+	−	+
Two orthologies	+	−	−
Gene phylogeny			
Paralogy	+	+	−
Xenology	−	+	+
Paraxenology	−	+	−
Plerology	−	+	−

genome. Gene duplication can occur after transfection and result in multiple copies of the xenologue. For instance, the mobile elements in *Drosophila* are probably of xenologous origin (Stacey *et al.* 1986), but as they occur in numbers are paraxenologous.

6.3.9 Plerology

A gene may be composed of a number of differing functional parts: exons which translate into proteins and correspond to structural, and sometimes functional, domains of proteins; and introns which do not code for amino acids and are excised prior to protein translation. Thus, if the order of exons and introns is changed within a single gene or between a number of other genes, new functions

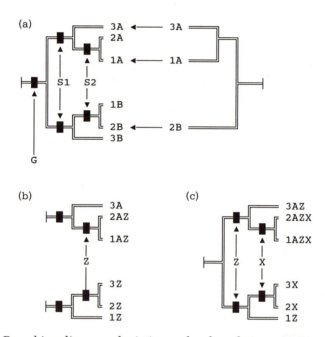

Fig. 6.2 Branching diagrams depicting molecular relations. (a) Diagrammatic representation of orthology and paralogy. A and B are different genes resulting from a gene duplication at point G prior to two speciation events, S1 and S2. If all genes are used to reconstruct the phylogeny, then genes A and B group together to give a correct species tree. If only some of the gene duplication products are used (as in the left-hand diagram) an incorrect tree can result. (b) Transfection of gene Z can influence the outcome of an analysis. Horizontal transfer is xenology. (c) Multiple transfection, as with genes X and Z, and gene conversion can result in a mixed gene product which may or may not coincide with the species phylogeny.

of proteins can occur along with new configurations between exons and introns. This process is known as gene conversion, and results in genes composed of bits (partial sequences) of other genes. This situation has led some to propose the idea of partial homology (for example Goodman *et al.* 1987). However, partial homology is based only on a consideration of similarity; the 'parts' will correspond to the genes of origin. The relation plerology (pleres = full, complete; Patterson 1988*a*; Fig. 6.2c) reveals a gene composed of 'parts' of other genes and will reconstruct a composite history of those genes.

It is clear from these relations that only orthologues can contribute to the resolution of species phylogenies directly and preliminary consideration should be given to establishing exactly the relation that the genes under consideration can discover.

6.4 NUCLEOTIDE SEQUENCE DATA: THE CHARACTERS

Using orthologous sequences the characters can be further examined. Character distinction is dependent upon two assumptions. First, characters are assumed to be independent, and second, that they show positional identity. The independence of characters is assumed in most character-based methods of analysis (Swofford and Olsen 1990). Yet in sequence data this may be an over-optimistic assumption (see later). Positional identity is established by aligning sequences to minimize mismatches; after alignment, sequences share either identical or different character states. For instance, in Table 6.3, characters 1, 2, and 4–8 have identical states for both species whereas characters 3 and 9 have different states. The explanation for non-identity is relatively simple: sites can differ by either a substitution event or an insertion/deletion (indel) event. The explanation of identity is more complex. Identity can either be primary identity (common ancestry, either synapomorphy or symplesiomorphy) or secondary identity (homoplasy: reversal, parallelism, convergence or chance similarity) (Patterson, 1988*a*). Distinguishing between these two types of identity is an analytical problem. Secondary identity (homoplasy) will be seen to cause most, if not all, the problems in establishing interrelationships between species.

Table 6.3 Hypothetical sequences for two species. Asterisks represent sites that differ

Position	1 2 3 4 5 6 7 8 9
Species A	AACGTTTAA
Species B	AAGGTTTAC
	* *

6.4.1 Substitutions

In nucleotide sequences, adenine (A) and guanine (G) are known as purines and cytosine (C) and thymine (T) as pyrimidines. Substitution events are divided into two kinds. Transitions are changes between either purines or pyrimidines, of which there are four possible options (Fig. 6.3a). Transversions are changes among purines and pyrimidines, of which there are eight possible options (Fig. 6.3b). Changes observed in aligned sequences can be thought of as an accumulation of transversional and transitional substitutions, where the frequency of these two classes of changes may be of different orders.

6.4.2 Gaps — insertions and deletions (indels)

It must be remembered that gaps are artefacts introduced by alignment programs and are not necessarily equivalent to insertions and deletions, which are mutational events (Olsen 1988). The addition of gaps into a set of sequences increases the number of characters and can change certain nucleotide associations: different alignments can suggest different characters if the procedure 'pushes' nucleotides into position with others it was not previously associated with. There are a number of methods for obtaining alignments. However, there is no guarantee that there is a single optimal alignment (Smith and Fitch 1981). Some current alignment programs use the dynamic programming approach of Needleman and Wunsch (1970), which maximizes matches between identical nucleotides. It assigns different scores to different matches so that it can distinguish between substitution events (mismatches) and indels (gaps). For example, a positive score (usually 1) is given to 'matches', that is identical nucleotides, a zero score to 'mismatches', that is putative substitution events as evidenced by different nucleotides and a negative score (termed a penalty) to allow the insertion of a gap. There are different ways of scoring. For instance, Sellers (1974) gives a positive score to 'mismatches' and gaps and an exact algorithm finds the alignment that minimizes the total distance between sequences.

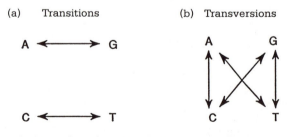

(a)   Transitions   (b) Transversions

Fig. 6.3 Classification of nucleotide substitutions, each arrow represents two options for direction of change. (a) Transitions. (b) Transversions.

It is now popular to calculate similarity scores between taxa based on some form of weighting (Wilbur and Lipman 1983). From these similarity scores a dendrogram is constructed using a simple clustering method such as UPGMA. The sequences are then aligned following the order specified by the dendrogram (Feng and Doolittle 1987; Higgins and Sharp 1988, 1989). The CLUSTAL program follows this protocol. This procedure highlights ways in which alignments and tree reconstruction methods can be considered simultaneously. There have been a number of recent studies examining this idea, recogized some years ago by Sankoff and his co-workers (Sankoff 1975; Konings *et al.* 1987; Hein 1989*a,b*; Nanney *et al.* 1989). It is important to note that many workers discard those areas of sequence matches that cannot be unambiguously aligned. It is generally acknowledged that visual inspection is the most widely used and probably considered the most reliable method (Swofford and Olsen 1990).

6.5 TREE CONSTRUCTION — PRELIMINARIES

In molecular systematics three types of tree diagram have been used to summarize results (Page 1990, fig. 5; Fig. 6.4). All of the diagrams include a representation of the branching order of taxa; two of the diagrams (Figs 6.4b and c) include information additional to branching order. A cladogram is a branching diagram depicting nested sets of synapomorphies resulting in a summary statement of sister-group relations among taxa — molecular systematists have used the term branching order, which may give the impression of a temporal sequence. However, it can be used in the context of topology position alone (Olsen 1988). A cladogram need not contain information about the relative distances between taxa at the terminal and internal nodes (Fig. 6.4a). To discover the branching order is the goal of parsimony procedures.

Branch length estimates, that is a measure of the distance between taxa and nodes, can be calculated and used to construct a tree. Branch lengths are said to express some notion of evolutionary distance between taxa by specifying a quantity between nodes. To construct trees using these principles requires additive data. To be additive the branch length distances can be added such that all pairwise distances are equal to the sum of the branch lengths that connect taxa (Swofford and Olsen 1990; Fig. 6.4b). Note that branch lengths leading to different taxa need not be equal and hence equality in the rates of evolution in any of the peripheral branches is not implied (Fitch 1984). This imposes a constraint on the tree: the data must be additive. Additivity is the basis for constructing trees using distance matrix methods.

A further restriction can be added to additive trees, namely that the branch lengths fit a tree so that the distance between any two taxa is equal to the sum of the branches joining them and all taxa are equidistant from the root (Swofford and Olsen 1990). The data are then ultrametric (Fig. 6.4c). This imposes a

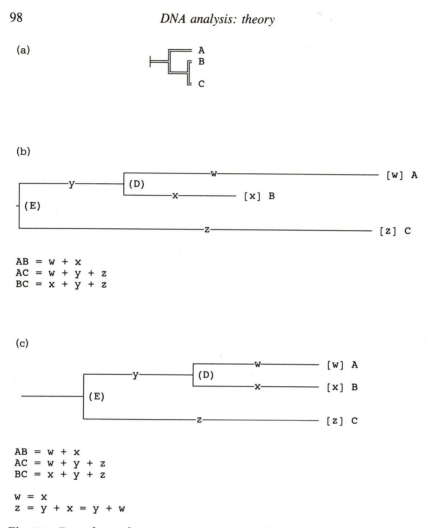

Fig. 6.4 Rooted trees for representing species relationships. (a) Cladogram. (b) Tree derived from additive distances; w–z are branch lengths; the equations hold for additivity. (c) Ultrametric tree; w–z are branch lengths; the equations hold for ultrametrics.

further constraint on the tree structure: first, the data must be additive (as ultrametric distances are additive) and secondly, all lineages diverge at an equal rate. This is equivalent to making the assumption of an exact molecular clock. Ultrametric trees are the basis for reconstructing trees using phenetic clustering measures such as UPGMA (Sneath and Sokal 1973).

Two of the trees in Fig. 6.4 can be represented as unrooted trees (or networks, Lundberg 1972; Swofford 1990; Fig. 6.5a,b). Of the three trees discussed

(a)

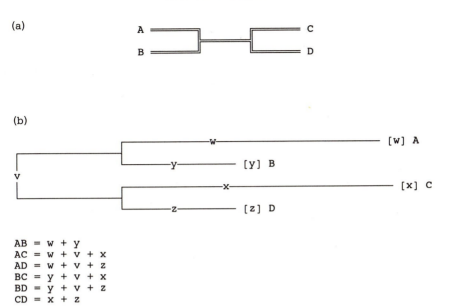

AB = w + y
AC = w + v + x
AD = w + v + z
BC = y + v + x
BD = y + v + z
CD = x + z

Fig. 6.5 Unrooted trees for representing species relationships. (a) Cladogram. (b) Tree derived from additive distances; w–z are branch lengths; the equations hold for additivity.

above, ultrametric trees cannot be represented as unrooted trees as, by definition, they must contain a root.

In Fig. 6.6 (data in Table 6.4) positions 4–12 support the tree ((A B)(C D)), positions 1–3 support the tree ((A C) (B D)) and positions 13 and 14 support the tree ((A D)(B C)). Some of these characters must be misleading, as all solutions cannot be correct. But which ones? The methods discussed below present various ways of dealing with this problem. Thus the primary concern in the determination of gene or taxon interrelationships is distinguishing between primary identity (homology) and secondary identity (homoplasy). Rooting the tree establishes the direction of character change (character polarity), distinguishes those character states which are synapomorphies and those which are symplesiomorphies, and which groups are monophyletic (see also Chapters 2 and 3). Rooting networks is accomplished by auxiliary methods (Farris 1972, 1979a; Fitch 1984).

6.6 NOTE ON ROOTING TREES

With four taxa there are three and only three possible unrooted trees and 15 fully bifurcating rooted trees (see also Chapters 1 and 2, Fig. 2.9; Lundberg 1972). Using the sequences given in Table 6.4 it can be seen that the states in positions

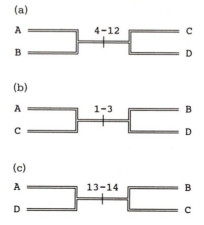

Fig. 6.6 Three possible solutions from the data in Table 6.5. Central branch indicates character number with support.

Table 6.4 Hypothetical sequences for four species. Solutions are given in Fig. 6.6

	1 1 1 1 1
Position	1 2 3 4 5 6 7 8 9 0 1 2 3 4
Species A	ATTAAGGCCCCCCG
Species B	GCGAAGGCCCCCAT
Species C	ATTTGAATTTTTAT
Species D	GCGTGAATTTTTCG

4–12, support the tree ((A B)(C D)). If, for the purpose of this example, we consider this the preferred solution, then there are five positions, one along each branch (four peripheral and one central branch), at which the tree can be rooted to give the five fully resolved cladograms.

The most common approach to rooting networks is outgroup comparison (Watrous and Wheeler 1981; Farris 1982a; Maddison *et al.* 1984; Darlu and Tassy 1987; see also Chapter 3). For simplicity only one character (Table 6.5) has been used which distinguishes two groups (A B) and (C D). Figure 6.7 illustrates the character configurations for each tree. In the tree of Fig. 6.7b both groups are monophyletic and both characters are synapomorphies, whereas in the other four topologies only one group is monophyletic and one character synapomorphic (excluding the group ABCD) (Fig. 6.7c–f). In an ideal situation the outgroup should be closely related to the ingroup; close enough to allow polarity to be established unequivocally.

Table 6.5 Hypothetical sequences for six species; only one site illustrated. Solutions are given in Fig. 6.7

Position	1 1 1 1 1
	1 2 3 4 5 6 7 8 9 0 1 2 3 4
Species A	A . . . .
Species B	A . . . .
Species C	T . . . .
Species D	T . . . .

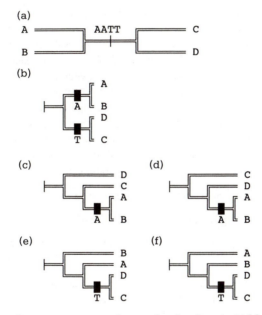

Fig. 6.7 (a) A solution, as unrooted tree, for the data in Table 6.6. (b–f) All possible configurations result from rooting the tree of Fig. 6.7a.

Without knowledge of the outgroup, an alternative procedure is midpoint rooting. The root of the tree is placed at the midpoint of the longest path connecting two lineages inside the study group (Farris 1972). Thus mid-point rooting is based not only on synapomorphies but also on the amount of difference between taxa. Midpoint rooting can only be successfully used if the two most divergent taxa can be shown to have the same rate of evolutionary change: a constant clock would have to be assumed (Farris 1972). This assumption has problems both theoretical (Farris 1981) and empirical (Britten 1986; Li and Tanimura 1987).

7.

DNA analysis: methods

David M. Williams

7.1 MATRIX METHODS

Distance matrix methods fit a tree to a matrix of pairwise distances between taxa (as represented by their genes). They are sometimes described simply as pairwise methods and are either wholly or conceptually related to phenetic methods (Sneath and Sokal 1973; Farris 1972, 1981; Felsenstein 1982; Fitch 1984). There should be an exact relationship between the observed distances and a unique tree. However, because the observed distances are approximations of actual distances they rarely fit a tree exactly (Fitch and Margoliash 1967; Felsenstein 1986). The best tree is that which most closely fits the observed to the tree distances and is evaluated by using some form of best-of-fit statistic (Swofford 1981). The construction of phylogenetic trees from sequence data using matrix methods can, but need not, involve three steps:

(1) conversion of a set of aligned sequences to a distance matrix;

(2) construction of topologies from those distances; and

(3) an evaluation of the best tree against all (or at least a selection of) other topologies.

Distance measures were first used to reconstruct phylogenetic trees on the amino acid sequences of cytochrome *c* (Fitch and Margoliash 1967). Their use with discrete characters, like amino acids and nucleotide sequences has continued. However, their suitability for such data has been questioned (Farris 1981, 1985). Yet some data, such as immunological and nucleic acid hybridization data, directly yield results in the form of distances and the need for such techniques may thus be required (Sarich 1969; Farris 1972; Springer and Krajewski 1989).

7.1.1 Definitions

It is important to clarify terminological usage, especially the distinction between similarity and distance (Swofford and Olsen 1990). Similarity is usually understood as a scale between two different parameters and expressed as a percentage

(0–100 per cent), such that identical objects are 100 per cent similar. Distance is sometimes treated as synonymous with dissimilarity, the direct opposite of similarity.

7.1.2 Distance measures

A distance measure is the degree of dissimilarity between two taxa or two genes. The simplest distance is the number of positions in which the pairwise comparisons differ. This can be expressed as a percentage or a fixed number. This measure is not a direct estimate of the number of substitutional events as it will be distorted by homoplasy – reversals, multiple-hits, and parallelisms. Without homoplasy, total evolutionary change will be equal to actual measured non-identity. Under such conditions distance measures will reconstruct a single tree as there is no ambiguity in the data. Yet is it exceedingly unlikely that real data contain no homoplasy. In fact, the absence of homoplasy is only evident after the observed distances are seen to exactly fit a unique tree. Thus even in this extremely unlikely situation a method is still required (Farris 1982*a*, 1983).

The initial pairwise measures are presented as a symmetrical taxon by taxon matrix which specify the distance between each pair of taxa.

7.1.3 Metrics

Additive data were briefly touched on above in the context of tree structure. How additivity influences tree construction can be illustrated with a simple example (adapted from Fitch 1981, 1984). In this example the 'true' tree is given in Fig. 7.1. The data are perfectly additive, such that the branch lengths can be added to the tree to sum exactly those observed. It should be noted that the most parsimoniously evolved topology will result if and only if no character changes its state more than once (Fitch 1981, 1984). Thus additivity obtains only if there is no homoplasy. In Fig. 7.2 the true tree is the same as that given in Fig. 7.1, but now the data include some homoplasy. New calculations give different sets of values for the branch lengths (matrix in Fig. 7.2). When the tree is reconstructed different values for the branch lengths are obtained which depart from their 'true' values. In this example the branching order remains the same; however, as conflicting evidence increases the branching order may eventually change.

Viewed in terms of evolutionary change a tree is perfectly additive if no homoplasy is present. In terms of tree construction algorithms, an additive tree is one that must satisfy four conditions (Dobson 1974; Sneath and Sokal 1973; Waterman *et al.* 1977; Rogers 1986). This can be summarized diagrammatically. In Fig. 7.3, A, B, and C are taxa and *d* is the distance between the pairs, four conditions are:

(1) $d(A,A) = 0$. There will be no measurable distance between the same taxon;

(2) $d(A,B) = d(B,A)$. The distances between taxa must be symmetrical. That
is the distance from A to B must involve the same quantity as from B to A;

(3) $d(A,B) \geq 0$. The distance must be non-negative; and

(4) $d(A,B) \leq d(A,C) + d(B,C)$. The triangle inequality must be satisfied.

Points (1) and (2) should be obvious and are non-contentious. Points (3) and (4)
are interrelated and require further explanation.

    Point (3) forbids negative branch lengths. Farris (1981) utilized this fact by
noting that negative branch lengths indicate non-metric distances. Furthermore,
he noted that branch lengths must be non-negative if the distance is to be proportional to amounts of evolutionary divergence (Farris 1981). Farris (1985)
discussed two ways in which non-negative distances can arise. First, the distance
is additive in expectation and the negative value is a sampling error (Nei *et al.*
1983; Felsenstein 1984); or else the data are non-additive in expectation and
the results indicate real non-additive distances. Homoplasy suggests that the
distances will be non-additive in expectation.

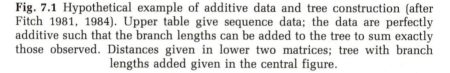

Fig. 7.1 Hypothetical example of additive data and tree construction (after
Fitch 1981, 1984). Upper table give sequence data; the data are perfectly
additive such that the branch lengths can be added to the tree to sum exactly
those observed. Distances given in lower two matrices; tree with branch
lengths added given in the central figure.

Species A	AAAAAAACCCCCCCCCCC
Species B	AAAAAAAAAAAAAAACCC
Species C	CCCAAAAAAAAAAAAAAA
Species D	CCCCCCCCCCCCAAAAAAA
Homoplasy	****

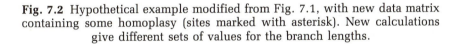

	Observed						Reconstructed			
	A	B	C	D			A	B	C	D
A	–					A	–			
B	8	–				B	8	–		
C	14	6	–			C	12	8	–	
D	14	14	8	–		D	16	12	8	–

Fig. 7.2 Hypothetical example modified from Fig. 7.1, with new data matrix containing some homoplasy (sites marked with asterisk). New calculations give different sets of values for the branch lengths.

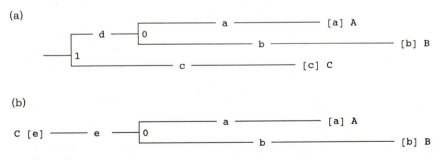

(a)

(b)

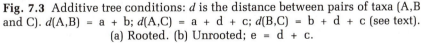

Fig. 7.3 Additive tree conditions: *d* is the distance between pairs of taxa (A,B and C). $d(A,B) = a + b$; $d(A,C) = a + d + c$; $d(B,C) = b + d + c$ (see text). (a) Rooted. (b) Unrooted; $e = d + c$.

Concerning point (4), the triangle inequality principle can be demonstrated with a simple example. Distances can be calculated as follows from the tree in Fig. 7.3:

$$d(A,B) = a + b; \; d(A,C) = a + d + c; \; d(B,C) = b + d + c.$$

Therefore:

$$d(A,B) + d(B,C) = a + d + c + 2b = d(AC) + 2b$$

As b cannot be negative (point (3) above), then:

$$d(A,B) + d(B,C) \geq d(A,C).$$

This allows each pairwise distance to be calculated. From these simple calculations it should be clear that in some instances negative distances may be required to satisfy the triangle inequality.

Phenetic clustering methods, like UPGMA, add a further restriction to those for additive tree construction. If the following three-point condition is satisfied, we have an ultrametric. For any three taxa, A, B, and C:

$$d(AC) \leq \max (d\text{AB}, d\text{BC})$$

An ultrametric requires the additional assumption that taxa are equidistant from each other and the root of the tree.

The way in which these conditions are used in tree construction will be examined below. The most restricted form is dealt with first, ultrametric trees and phenetic clustering; this will be followed by distance measures and additive trees.

7.1.4 Tree reconstruction — phenetic methods

Ultrametric distances behave as if a precise molecular clock exists by forcing the assumption of equal rates of change in all lineages because ultrametric data require the assumption that taxa are equidistant from each other and the root of the tree. There is no guarantee that amount of divergence is exactly linear with time and ultrametric data are highly unlikely − even more unlikely than additive data. The justification for equal rates of evolution arose from consideration of the neutral theory of evolution (Kimura 1983). However, there is a considerable body of empirical evidence that suggests equal rates of evolution, in many instances, do not occur (Britten 1986; Hillis 1987; Wheeler and Honeycutt 1988; Hillis and Moritz 1990). Although it is recognized that a general relationship between divergence and time exists, it may be necessary to calibrate each protein clock separately rather than assuming equal rates prior to tree reconstruction (Scherer 1990). However, if it is possible to show that the data are ultrametric then a simple cluster analysis can be used (see Page 1990, 1991 for examples).

UPGMA utilizes these principles, calculating a tree on a single pass of the data (Sneath and Sokal 1973). There are many versions of the UPGMA program, one of which is available in Nei's package of programs while a more sophisticated version is in the NYTSYS package (Rohlf 1988). The program KITSCH in the PHYLIP package (Felsenstein 1989), a variant on the Fitch−Margoliash algorithm (see later), constrains the tree to assume equal rates of change on all lineages, thus assuming a molecular clock.

7.1.5 Tree reconstruction — distance methods

If the properties of additive trees are utilized rate heterogeneity is not a problem. This is easily demonstrated by observing that the tree in Fig. 7.4 has branches with differing lengths, indicating different rates. The first method to utilize the distance matrix method was described by Fitch and Margoliash (FM; 1967). They used an additive branch length fitting technique in which trees are constructed with branches associated with a numerical length. First, a matrix of distances between pairs of terminal taxa is produced. The tree distance for the pair is the sum of the lengths of their branches. When compared these measures may be different and the difference is minimized in an attempt to gain a best fit. The tree is constructed by first joining together two taxa that are most similar in the matrix. It is assumed that the pair of taxa that have the smallest distance are most closely related. This shows the conceptual relationship between their method and phenetic methods, which proceed in the same fashion. Once joined, the two taxa are combined and are then joined to the next shortest distance. This process is continued until all taxa are joined (Fitch 1977).

Thus far, this method closely approaches that of phenetic clustering techniques by selecting the two taxa with the smallest distance for the first union (Sneath and Sokal 1973). However, subsequent rearrangements of the primary topology can be made in an attempt to find other trees which may better fit the observed distances. The fit needs to be assessed. In the FM method %SD (standard deviation) is used. This is a weighted least square estimate of branch lengths. Thus, the best tree is the one in which the reconstructed distances match most closely the measured distances. Standard deviation is a common way to assess the closeness of data pairs (Fitch and Margoliash 1967; Fitch 1977). For a perfect fit this will be zero (see example in Figs 7.5 and 7.6) and goodness of fit is determined by how far from zero this measure deviates (Tables 7.1 and 7.2). However, since there are a large number of trees to check, and although Swofford (1981) has suggested using tree rearrangement techniques, it is still difficult to be certain of discovering the tree(s) with the lowest %SD. To compensate for this some algorithms have been constructed in order to combine the %SD measure with the construction of additive trees (De Soete 1983); or to incorporate rearrangement techniques at early stages of the tree-building process (Tateno 1990).

A number of problems exist with distance methods, most of which are concerned with measuring the distance between taxa and what those measurements mean. Beyer *et al.* (1974) first criticized the FM method because a number of distances resulted in negative lengths and failed to satisfy the triangle inequality. Although Felsenstein modified the FITCH program in the PHYLIP package to avoid negative branches (an option is available in the FITCH program that will retain negative branch lengths if they exist), Farris' (1985) point remains: negative branch lengths identify non-additive data and removing this aspect avoids empirically testing this objective (for example Carpenter 1990).

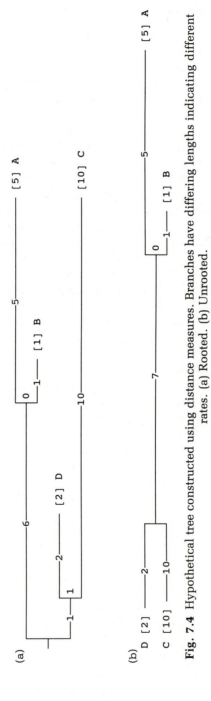

Fig. 7.4 Hypothetical tree constructed using distance measures. Branches have differing lengths indicating different rates. (a) Rooted. (b) Unrooted.

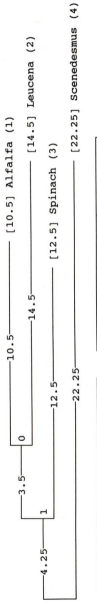

                                                        [10.5] Alfalfa (1)
                                                        [14.5] Leucena (2)
                                                        [12.5] Spinach (3)
                                                        [22.25] Scenedesmus (4)

Observed

	1	2	3	4
1	–			
2	25.0	–		
3	26.5	30.5	–	
4	40.5	44.5	39.0	–

Reconstructed

	1	2	3	4
1	–			
2	25.0	–		
3	27.0	30.0	–	
4	40.0	45.0	39.0	–

Fig. 7.5 Weighted least square estimate of branch lengths from example given in Fitch (1977).

Observed				
	1	2	3	4
1	–			
2	25.0	–		
3	26.5	30.5	–	
4	40.5	44.5	39.0	–

Reconstructed				
	1	2	3	4
1	–			
2	25.0	–		
3	26.5	30.5	–	
4	40.5	44.5	39.0	–

Fig. 7.6 Weighted least square estimates of branch length will be zero for a perfect fit.

Table 7.1 Calculations of weighted least square fits for distances given in Fig. 7.5 (After Fitch 1977)

$$\%SD = 100 \sqrt{\sum_{ij}} \frac{[d(i,j) - p(i,j)]}{d(i,j)}$$

$$
\begin{aligned}
d(i,j) - p(i,j) = d(A,B) &= 25.0 - 25.0 = 0 \\
= d(A,C) &= 27.0 - 26.5 = 0.5 \\
= d(A,D) &= 40.5 - 40.0 = 0.5 \\
= d(B,C) &= 30.5 - 30.0 = 0.5 \\
= d(B,D) &= 45.0 - 44.5 = 0.5 \\
= d(C,D) &= 39.0 - 39.0 = 0
\end{aligned}
$$

times 100, 0.5 becomes 50; then

$$
\begin{aligned}
/d(i,j) = d(A,B) &= 0 \\
= d(A,C) &= 50/27.0 = 1.85 \times 1.85 = 3.43 \\
= d(A,D) &= 50/40.0 = 1.25 \times 1.25 = 1.56 \\
= d(B,C) &= 50/30.0 = 1.67 \times 1.67 = 2.78 \\
= d(B,D) &= 50/45.0 = 1.11 \times 1.11 = 1.23 \\
= d(C,D) &= 0
\end{aligned}
$$

Sum of squares is
$3.43 + 1.56 + 2.78 + 1.23 = 9.00$
divided by $ij\ (= 5) = 1.8$, the square root = **1.34**

Table 7.2 Calculations of weighted least square fits for distances when a perfect fit is achieved. (After Fitch 1977)

$$
\begin{aligned}
d(ij) - p(i,j) = d(A,B) &= 25.0 - 25.0 = 0 \\
= d(A,C) &= 26.5 - 26.5 = 0 \\
= d(A,D) &= 40.5 - 40.5 = 0 \\
= d(B,C) &= 30.5 - 30.5 = 0 \\
= d(B,D) &= 44.5 - 44.5 = 0 \\
= d(C,D) &= 39.0 - 39.0 = 0
\end{aligned}
$$

Farris' distance Wagner method (Farris 1972; Swofford 1981) is an improvement upon the FM as it does not allow negative values for branch lengths. However, this is not by prohibition but by restricting acceptable tree topologies only to those that have non-negative branch lengths. Farris' original method also proceeds by fitting expected to observed distances, leading to the construction of a tree by the progressive addition of new taxa (Farris 1972). Farris' modification constrained the observed distances to be either greater than or equal to the tree distances. Hence, the sum of all differences between observed and tree distances is minimized, such that the algorithm attempts to fit 'true' (that is evolutionary) and input distances (Farris 1972; Faith 1985).

The distance-Wagner algorithm differs from the character-Wagner algorithm in that the latter can assign character-states to the new nodes (or HTUs; Farris 1970). With only a matrix of pairwise distances this is not possible and these values must be estimated. The practical aspect of this procedure has been discussed and modifications have been suggested by various authors (Tateno *et al.* 1981; Faith 1985).

7.1.6 Problems

The key problem with distance matrix methods is non-additivity. A number of methods have been devised to circumvent the problem of non-additive distances, either by relaxing the criterion for additivity or transforming the distances prior to analysis, modifying them to more closely resemble 'evolutionary' distances. A combination of the two can be used, and may be preferred (Saitou and Nei 1987). Given the appropriate conditions, the original distances could be transformed into additive distances prior to analysis (Li 1981; Klotz *et al.* 1979; Klotz and Blanken 1981). With distances appropriately corrected a simple clustering procedure can be used to construct the tree. What, then, are the appropriate conditions? A number of procedures have been suggested to compensate for the effects of homoplasy (and, as a consequence, non-additivity) to produce evolutionary distances (Olsen 1988; Swofford and Olsen 1990) or transformed sequence distances (Felsenstein 1984).

To assess the impact of homoplasy some knowledge of the pattern of homoplastic change is necessary, such that the relationship between total divergence and homoplasy can be explored. There are two options available. If the possibility that substitutions accrue in a probabilistic manner at all positions and in all lineages is considered, then the expected number of homoplastic sites between two taxa is a function of their substitution rates. Thus, as divergence increases so does the frequency of homoplasy and, given enough time, all signal will be lost. The problem would be to discover at what evolutionary distances the molecule remains useful (see, for example, Smith 1989; Miyamoto and Boyle 1989). If, on the other hand, the accumulation of homoplasy is stochastic then it will be almost impossible to estimate substitution rates prior to tree reconstruction. The former model is supported by empirical observations

(Thorpe 1982) and solutions have been sought in various probabilistic models of the evolutionary process (Zuckerkandl and Pauling 1965; Kimura 1969; Jukes and Cantor 1969; Holmquist 1972; Tajima and Nei 1984; summarized in Bishop and Friday 1985).

Probabilistic models were originally based on a simple Poisson process applicable to every site with back mutations ignored (Jukes and Cantor 1969). Yet, it was quickly realized that pairwise substitution probabilities among bases can be unequal. For example, the frequently acknowledged transition/transversion bias observed in many species (for example Brown *et al.* 1982). Kimura's '2-parameter' model allows for unequal rates between transition and transversion rates (Kimura 1980) and is available in the PHYLIP package.

In addition, substitution probabilities among positions can also be unequal. Base pairing reduces the number of independent samples such that they can introduce unnecessary noise to the data. Weighting procedure have been suggested for the imbalance observed between paired and unpaired site differences (Olsen 1988; Wheeler and Honeycutt 1988).

For sequences that code for amino acids there are further complicating factors which can only be touched upon here. For example, due to the degeneracy of the genetic code most amino acids are encoded by several synonymous codons. In addition, biased codon usage affects substitution probabilities in functional genes. Calculations can be made to effect particular weighting schemes (for example Fitch and Markowitz 1970).

After such adjustments, the data are said to be closer to 'real' (or actual) evolutionary distances. The development of these types of evolutionary models for phylogeny reconstruction has become prominent in maximum likelihood methods, yet they deal with only a few taxa and very simplistic models (Felsenstein 1981*b*; Bishop and Friday 1985; see also Chapter 5).

An alternative way of dealing with non-additivity is to relax the definition by forming clusters consistent with the largest fraction of taxon quartets. Methods of this sort are additive similarity trees (Sattath and Tversky 1977), the method of Fitch (1981) and, of a different kind, the neighbour-joining method (Saitou and Nei 1987; Studier and Kepler 1988).

Fitch (1981) and Sattath and Tversky (1977) both considered tree topologies in terms of neighbouring pairs, where neighbours are defined as two taxa separated by a single node and two branches. The method requires searches between groups of four taxa in an effort to find the additive condition between neighbouring sets. The relaxed additive condition will be:

$$d(A,B) + d(C,D) \le d(A,C) + d(B,D) \le d(A,D) + d(B,C).$$

Thus for any four taxa and their six associated pairwise distances, the two pairs of taxa that are considered as neighbours satisfy the above four point condition even if additivity is approximate (Fitch 1981). In Fig. 7.7, if $d(A,B) < d(A,D)$, and $< d(A,C)$, then A, B and C, D are two pairs of neighbours.

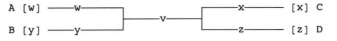

Fig. 7.7 Diagrammatic representation of relaxed additivity used in neighbourliness methods. If $d(A,B) < d(A,D)$ and $< d(A,C)$ then A,B and C,D are two pairs of neighbours.

In conclusion, distance analyses suffer because probably all sequence data are non-additive (Farris 1981). As stated above, non-additive distances can be identified by the presence of negative branch lengths (Farris 1981). Negative branch lengths are methodological artifacts as they clearly cannot be proportional to evolutionary divergence. Felsenstein (1986) has turned this point by stating that distances are additive (that is non-negative) in expectation. That is, they are only estimates. Yet as branch lengths increase, the accuracy of the estimate decreases (Farris 1986*b*). It is not clear what the distances represent and this has led to either abandoning distance techniques altogether or using only those methods that are truly metric.

7.2 PARSIMONY METHODS

Parsimony methods attempt to find the tree that requires the least number of changes to explain the observed data, in this case nucleotide sequences (Farris 1983; Felsenstein 1982, 1988*b*). Parsimony methods rely on minimizing the number of steps for the transformation of one character to another and conflicts in the data are minimized (Farris 1983). This is established by measuring the tree length.

There are three stages: the optimality criterion used to infer the tree, the algorithm that is employed in the search for optimal trees under those conditions, and the measure used to evaluate the result (see also Chapter 3; Swofford and Olsen 1990). The optimality criterion is that which specifies the restrictions imposed on character changes, while the algorithm that searches for the optimal tree is concerned with the practical problem of inspecting all possible topologies.

All the above have been discussed in detail earlier (see also Chapter 4) and so will be treated here only for the special case of nucleotide sequence data.

Camin and Sokal (1965) first proposed a discrete character parsimony analysis. Their method included the restriction that character transformations are irreversible, that is they assumed evolution to be irreversible. It is obvious that such a restriction will be unwarranted in most, if not all, cases and this condition has since been relaxed.

The Wagner method is appropriate for binary or ordered multistate characters (Kluge and Farris 1969; Farris 1970; Farris *et al.* 1970). Yet it is unlikely that there would be sufficient knowledge to allow ordering of nucleotide character states.

Fitch (1971) expanded the original Wagner algorithm to allow for unordered multistate characters, where any state is allowed to transform directly into any other state; that is any nucleotide can transform with equal probability into any of the others (see also Hartigan (1972) for the proof of Fitch's procedure). This is the most frequently used initial condition and is equivalent to uniformly weighted parsimony of Williams and Fitch (1989, 1990). Parsimony analyses allow the possibility of distinguising between potentially informative and uninformative sites. In Fig. 7.8, character 1 has the same base in each species and no changes are required to account for it on any of the trees, hence it is uninformative on the interrelations of these species. Character 2 has only one base changed in one species. If this is placed on each of the three alternative trees one step is required to explain its occurrence regardless of the tree; hence it too is uninformative. The same situation is observed with character 3, which requires two changes on every tree. The situation is different for characters 5

Position	12345678
Species A	CCTCAGTA
Species B	CTAAAGAA
Species C	CTCGGGAA
Species D	CTCTGGTA
	**

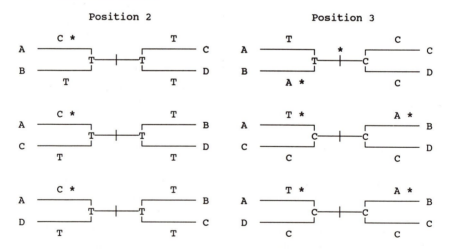

Fig. 7.8 Demonstration of the parsimonious distribution of character states on alternative possible trees. Character 1 has the same base at the same position in each species and no changes are required to account for it on any of the trees hence it is uninformative. Character 2 has one position changed in one species. If this is placed on each of the three alternative trees one step is required to explain its occurrence regardless of the tree; it too is uninformative; likewise character 3 which requires two changes on every tree.

and 8, which are potentially informative but support different trees; in these instances the competing trees need two changes rather than one (Fig. 7.9). The significant, informative changes occur in the central branch; the branch that divides two pairs of taxa (Fig. 7.9). Thus it is only the informative sites which can support one of the three solutions unambiguously and the tree which has the greatest central branch support is the one which is favoured in the maximum parsimony approach.

7.2.1 Weighting

In view of empirically observed bias in sequence data (such as transition/transversion bias), uniform weighting is considered by many as far too relaxed an assumption and as a consequence could result in either an overwhelming array of equally parsimonious topologies or little resolution because of the inclusion of misleading characters (Wheeler 1990*b*).

Position	12345678
Species A	CCTCAGTC
Species B	CTAAAGCT
Species C	CTCGGGCC
Species D	CTCTGGTT
	* *

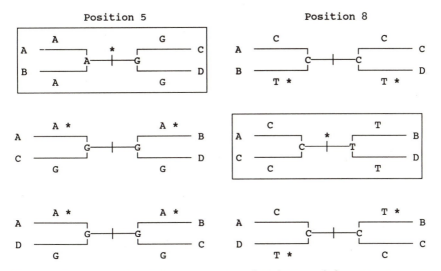

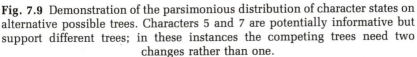

Fig. 7.9 Demonstration of the parsimonious distribution of character states on alternative possible trees. Characters 5 and 7 are potentially informative but support different trees; in these instances the competing trees need two changes rather than one.

What may be required is an evaluation of the characters used. One option is to weight certain characters based upon empirically observed frequencies in the data. By recognizing that characters can be weighted it follows that some are considered more informative than others in that they direct the result towards the 'correct' tree, while other characters are misleading as they direct the result towards 'incorrect' trees. But without the benefit of prior knowledge, how can it be decided which characters are useful and which are not? To return to an observation made earlier: useful characters are those that group taxa. These are based upon shared similarity at site positions. Yet, as was seen earlier, shared similarity can be explained either by primary (homology) or secondary (homoplasy) identity. In one view, then, if shared similarities as putative synapomorphies are the only evidence, some way of evaluating possible topologies is required. This approach is exemplified in Lake's linear invariant method (Lake 1987*a,b*). In another view, characters could be weighted prior to analysis to reflect any empirical bias seen in the data or by their performance on the initial reconstructed topology. In either view the data themselves need to be further examined in order to establish their 'quality'.

Weighting characters can be achieved either prior to tree construction and reflect what may be known of the data, or after tree construction and reflect what contribution characters make to the resulting topology (Neff 1986; Sharkey 1989). The two approaches have been called a priori and a posteriori weighting by Neff (1986), or hypothesis dependent and independent weighting by Sharkey (1989).

7.2.2 A priori weighting

Uniformly weighted parsimony has already been discussed as a standard where the initial estimate would be to assign equal weight to all positions used in the analysis (Williams and Fitch 1990). However, weights could be assigned before analysis in a number of ways.

Character weighting can apply to a particular position in an aligned sequence, such that greater favour is given to that site over all other sites in a proportional amount. Thus in Table 7.3a a weight is assigned to characters 4 and 8 five times that assigned to all other characters. Examples of this are the differential weighting of paired to unpaired sites; where paired sites are those which form Crick–Watson base pairs in the stem regions of a molecule and unpaired sites are those that occur in the loop regions of a molecule (Wheeler and Honeycutt 1988; Smith 1989; Fig. 7.10). A second example is the observed higher frequency of transitions to transversions (Brown *et al.* 1982), where greater weight may be given to transversions as they change more slowly. Transitional changes may become saturated and uninformative even at small divergence times (for example Miyamoto and Boyle 1989). These two types of weighting are actually fundamentally different, in that the first method weights entire characters while the second method weights parts of characters, i.e. some transformations receive

Table 7.3 Uniform and differential weighting of transversions. (a) Weights assigned to characters 4 and 8 are five times that assigned to all other characters. (b) Weight of 5 assigned to the transversion C → G (wherever it occurs in the sequence) and the transversion A → C assigned a weight of 8 (wherever it occurs in the sequence)

	a	b
Weight	1 1 1 5 1 1 1 5 1	1 1 5 1 1 1 1 1 8
Position	1 2 3 4 5 6 7 8 9	1 2 3 4 5 6 7 8 9
Species A	AAGCTTTAA	AACGTTTAA
Species B	AAGGTTTCA	AAGGTTTAC

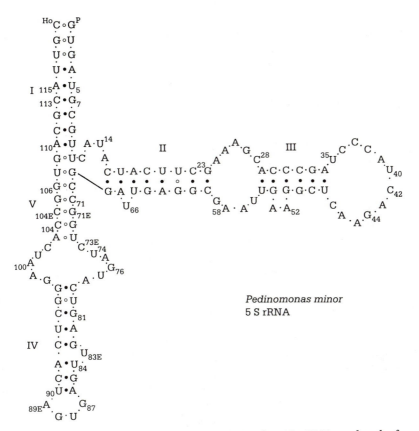

Fig. 7.10 Example of secondary structure in the 5S rRNA molecule from *Pedinomonas* (after Devereux *et al.* 1990). Paired sites are those which form Crick–Watson base pairs in the stem regions of a molecule and unpaired sites are those that occur in the loop regions.

higher weights than others. It is possible to apply both at ' once to the same sequence (Table 7.3). An extreme version of this weighting method is transversion parsimony which effectively recodes the sequence data into purines and pyrimidines and down weights evidence from transitional mutations completely. In general, a weight could be given to a particular character state transformation such that, for instance, the transformation from $C \rightarrow G$ will differ from $A \rightarrow C$. In Table 7.3b transformation $C \rightarrow G$ has a weight of 5 (wherever the transformation occurs on the sequence) and $A \rightarrow C$ has a weight of 8 (wherever the transformation occurs on the sequence). This has been called existential weighting (Williams and Fitch 1989, 1990).

With the above in mind, it can be seen that different optimization procedures utilize different kinds of weighting. Swofford and Olsen (1990) have suggested a way that these can be represented in a matrix which assigns different 'costs' to particular strategies (see also Chapter 3). They call this approach generalized parsimony (Swofford and Olsen 1990) based upon the construction of step matrices first discussed some years ago by Sankoff (1975). However, such calculations can be computationally exhaustive and often impractical. Neverthless, PAUP v.3.0 has this facility.

7.2.3 A posteriori weighting

Weights can be assigned to characters by their performance in tree construction. These methods require an initial tree. Examples of such procedures are successive approximation weighting (Farris 1969), quadratic weighting (Fitch and Yasunobu 1975), transformation series analysis (Mickevich 1982), and dynamic weighting (Williams and Fitch 1989).

As shown in Chapter 5, Farris' (1969) method sets weights according to their fit on the most parsimonious trees, giving higher weight to the most consistent characters. This method bears some relation to compatibility methods in that it supports characters that are consistent with the initial topology (Felsenstein 1981a; West and Faith 1990).

7.2.4 Problems

Parsimony analyses have been criticized because errors in the branching order may occur due to unequal rates of evolution (Felsenstein 1978b). Although parsimony methods do not assume constant rates, problems can occur under certain circumstances. When two (or more) branches undergo extensive substitution after taxa diverge, the changes in these 'long' branches may display many parallel changes which provide support for the 'wrong' tree (Felsenstein 1978b). Felsenstein (1978b) and Lake (1987a) suggested that this was caused by unequal rates of substitution and would lead parsimony methods to increasingly yield incorrect trees as data accumulates. However, the conditions under which parsimony fails to identify the correct tree have been further studied and the conclusion is that it is the presence of long branches, not the relative rates of substitution, that cause the selection of the wrong tree; the wrong tree can be selected by parsimony even with equal rates of substitution (Hendy and Penny 1989). This conclusion would point to the data as the source of difficulty not the method, as one way to circumvent this is to sample taxa which may 'break up' the long branches. Yet these taxa may be difficult to identify or may not be available.

The problem lies in the detection of misleading from informative patterns.

7.2.5 Invariants

Lake's method of evolutionary parsimony was specifically formulated to over-come the problem of differential rates in parsimony analyses (Lake 1987*a*). The method was initially outlined for dealing with four sequences. Recently, Cavender (1989) has extended it in principle to more sequences but it has yet to be applied to real data.

Lake utilized the fact that with four taxa (or sequences) there are only three possible unrooted topologies (Fig. 7.11). Support for each topology is retrieved from the record of transversional changes in the central branch, that is the characters which distinguish pairs of taxa. Lake used transversional substitutions as central branch evidence as they are considered more reliable. Lake (1987*a*) further distinguished between transversional changes that occur in the central branch of the tree from those that may occur in the peripheral branches; that is changes occurring after divergence. This can be demonstrated by a simple example. In Fig. 7.12 the initial configuration supports an ((A B)(C D)) tree.

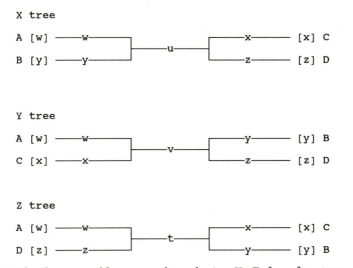

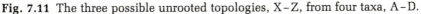

Fig. 7.11 The three possible unrooted topologies, X–Z, from four taxa, A–D.

If two further substitutions are assumed then the configuration becomes support for the ((A D) (B C)) tree. Lake noted that a correction could be made as there are two possible transversions that would occur in equal number for the incorrect trees and should statistically approach zero for the wrong trees. This embodies the principle of balanced transversions (Li and Nei 1990; Sidow and Wilson 1990); that is, a G → C transformation is considered as likely as a G → T. Lake developed formulae that permit computation of the relative support for each tree, selecting the tree that requires the minimum number of consistent substitutions; that is consistent with the substitution events that have occurred in the peripheral branches.

Lake's formulation depends upon three assumptions: that all substitution events are independent, equally likely and insertion and deletion events can be safely ignored. More significantly, the method holds if the assumption of balanced transversions is maintained (Li and Nei 1990; Sidow and Wilson 1990). Empirical evidence suggests that transversions are not always balanced (Nei 1987). However, this can be allowed for by calculating base composition frequencies for each class of change (Sidow and Wilson 1990). To appreciate the implication of Lake's formulae the type of character information and how they are used in parsimony methods needs to be understood.

Lake (1987*a,b*) used a system of vector representation for all possible patterns that can be produced from the set of four aligned sequences. Starting from sequence 1, all nucleotides that are the same in sequences 2 to 4 are assigned the code 1; those related by a transition are given a 2 and the two possible transversions are given a 3 and 4 in the order that they are encountered, if at all. Identical patterns will have the same notation, thus if the characters at two different positions are AAAA and GGGG both are recorded as 1111 and both contribute the same kind of information. This reduces the 256 possible permutations (Table 7.4) to 36 (Table 7.5). Thus, of the 256 possible types of characters, there are, in fact, only 36 possible configurations, or pattern types

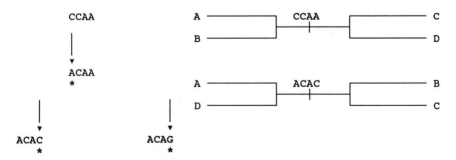

Fig. 7.12 Changes in sequences from support for one configuration to another. The initial configuration supports an ((A B)(C D)) tree. If two further substitutions occur then the configuration becomes support for the ((A D)(B C)) tree.

Table 7.4 There are 256 possible permutations of our sequences; only 36 patterns exist in the 256 permutations. Not all permutations are illustrated here

AAAA	1111	O	CAAA	1333	A	GAAA	1222	a	UAAA	1333	A
AAAC	1113	D	CAAC	1331	G	GAAC	1223	q	UAAC	1332	Q
AAAG	1112	d	CAAG	1332	Q	GAAG	1221	g	UAAG	1334	r
AAAU	1113	D	CAAU	1334	r	GAAU	1223	q	UAAU	1331	G
AACA	1131	C	CACA	1313	F	GACA	1232	l	UACA	1323	L
AACC	1133	E	CACC	1311	B	GACC	1233	H	UACC	1322	h
AACG	1134	J	CACG	1314	N	GACG	1231	s	UACG	1324	v
AACU	1132	k	CACU	1312	p	GACU	1234	u	UACU	1321	t
AAGA	1121	c	CAGA	1343	m	GAGA	1212	f	UAGA	1343	m
AAGC	1123	j	CAGC	1341	S	GAGC	1213	n	UAGC	1342	w
AAGG	1122	e	CAGG	1322	h	GAGG	1211	b	UAGG	1344	i
AAGU	1123	j	CAGU	1342	w	GAGU	1213	n	UAGU	1341	S
AAUA	1131	C	CAUA	1323	L	GAUA	1232	l	UAUA	1313	F
AAUC	1134	J	CAUC	1321	t	GAUC	1234	u	UAUC	1312	p
AAUG	1132	k	CAUG	1324	v	GAUG	1231	s	UAUG	1314	N
AAUU	1133	E	CAUU	1344	i	GAUU	1233	H	UAUU	1311	B
ACAA	1311	B	CCAA	1133	E	GCAA	1322	h	UCAA	1233	H
ACAC	1313	F	CCAC	1131	C	GCAC	1323	L	UCAC	1232	l
ACAG	1312	p	CCAG	1132	k	GCAG	1321	t	UCAG	1234	u
ACAU	1314	N	CCAU	1134	J	GCAU	1324	v	UCAU	1231	s
ACCA	1331	G	CCCA	1113	D	GCCA	1334	r	UCCA	1223	q
ACCC	1333	A	CCCC	1111	O	GCCC	1333	A	UCCC	1222	a
ACCG	1332	Q	CCCG	1113	D	GCCG	1331	G	UCCG	1223	q
ACCU	1334	r	CCCU	1112	d	GCCU	1334	r	UCCU	1221	g
ACGA	1321	t	CCGA	1134	J	GCGA	1312	p	UCGA	1234	u
ACGC	1323	L	CCGC	1131	C	GCGC	1313	F	UCGC	1232	l
ACGG	1322	h	CCGG	1133	E	GCGG	1311	B	UCGG	1233	H
ACGU	1324	v	CCGU	1132	k	GCGU	1314	N	UCGU	1231	s
ACUA	1341	S	CCUA	1123	j	GCUA	1342	w	UCUA	1213	n
ACUC	1343	m	CCUC	1121	c	GCUC	1343	m	UCUC	1212	f
ACUG	1342	w	CCUG	1123	j	GCUG	1341	S	UCUG	1213	n
ACUU	1344	i	CCUU	1122	e	GCUU	1344	i	UCUU	1211	b
AGAA	1211	b	CGAA	1344	i	GGAA	1122	e	UGAA	1344	i
AGAC	1213	n	CGAC	1341	S	GGAC	1123	j	UGAC	1342	w
AGAG	1212	f	CGAG	1343	m	GGAG	1121	c	UGAG	1343	m
AGAU	1213	n	CGAU	1342	w	GGAU	1123	j	UGAU	1341	S
AGCA	1231	s	CGCA	1314	N	GGCA	1132	k	UGCA	1324	v
AGCC	1233	H	CGCC	1311	B	GGCC	1133	E	UGCC	1322	h
AGCG	1232	l	CGCG	1313	F	GGCG	1131	C	UGCG	1323	L
. . . .	....	...	. . . .	....	...	. . . .	....	...	. . . .	....	...
AUUC	1334	r	CUUC	1221	g	GUUC	1334	r	UUUC	1112	d
AUUG	1332	Q	CUUG	1223	q	GUUG	1331	G	UUUG	1113	D
AUUU	1333	A	CUUU	1222	a	GUUU	1333	A	UUUU	1111	O

Table 7.5 (a) The 36 patterns of the 256 possible permutations of Table 7.4, and Lake's letter notation for the 36 pattern types. (b) Formulae for support using maximum parsimony, transversional parsimony, 'neighbor-joining' method, transversion 'neighbor-joining' and evolutionary parsimony

(a)

1111	O	1233	H
1112	d	1234	u
1113	D	1311	B
1121	c	1312	p
1122	e	1313	F
1123	j	1314	N
1131	C	1321	t
1132	k	1322	h
1133	E	1323	L
1134	J	1324	v
1211	b	1331	G
1212	f	1332	Q
1213	n	1333	A
1221	g	1334	r
1222	a	1341	S
1223	q	1342	w
1231	s	1343	m
1232	l	1344	i

(b)

Maximum parsimony
$X = e + E$
$Y = f + F$
$Z = g + G$

Transversion parsimony
$X = E$
$Y = F$
$Z = G$

'Neighbor-joining'
$X = 2e + 2E + H + h + i + J + j + k$
$Y = 2f + 2F + L + l + m + N + n + p$
$Z = 2g + 2G + Q + q + r + S + s + t$

Transversion 'neighbor-joining'
$Y = E + H + J + u$
$Y = F + L + N + v$
$Z = G + Q + S + w$

Evolutionary parsimony
$X = E - H - J + u$
$Y = F - L - N + v$
$Z = G - Q - S + w$

(Felsenstein 1991), and not all of these configurations are informative. In addition, Lake assigned a letter label to each of these 36 patterns. Thus the pattern 1111 was given the letter O (Table 7.5): all possible permutations can be represented by a single letter. Lake's system allows a notation view of which configurations are used to generate support for different methodologies. Returning to the three possible trees from four taxa and referring just to the X tree (Fig. 7.11), maximum parsimony uses two terms of support: E + e, that is all the transversional and transitional changes that occur in the central branch. For transversional parsimony only one term of support is used: E, all of the transversional changes (Table 7.5).

However, Lake's formulae utilize four terms, two of which provide support, E and u, and two which provide counter support, H and J, thus allowing changes that may mimic parsimony terms to be taken into account. Interestingly, if sites are coded into pyrimidines and purines then the counter support becomes support (Swofford and Olsen 1990; West and Faith 1990) and is equivalent to the transversion 'neighbor-joining' distance method (Li *et al.* 1987; Table 7.5). Lake's method has been discussed by a number of different authors (Penny 1988; Felsenstein 1988*c*; Patterson 1989) and appears to be devoid of the pitfalls of other methods: it does not require prior estimates on inferred branch lengths, constancy of rate and knowledge of the transition/transversion bias. However, reservations over its use have been expressed (Li and Nei 1990; Sidow and Wilson 1990) and the implications behind the formulae are being actively explored (Felsenstein 1991). Nevertheless, it does highlight the need for detailed knowledge of the characters in a set of sequences and what effect they are playing in the analysis.

7.3 SUMMARY

Parsimony methods have a distinct advantage as nucleotide sequences are in the form of discrete character data and the specific sites which contribute to systematic grouping can be identified (Swofford and Olsen 1990). It appears that, rather than explorations of methodology, examination of the sequences of particular genes are required, to determine just what phylogenetic information is contained within and what use it may be in the reconstruction of relationships (see, for example, the conclusions concerning the use of the 5S rRNA in Steele *et al.* 1991).

8.
Fossils and cladistic analysis

Peter L. Forey

8.1 INTRODUCTION

The problems and advantages of including fossils in cladistic analysis are discussed in this chapter. There is a limited literature on the subject and for present purposes the concepts are organized as follows: fossils and ancestors, age and rank, stem and crown groups, and the influence of fossils on classifications of Recent organisms.

8.2 FOSSILS AND ANCESTORS

Cladistic analysis is most commonly used to reconstruct the phylogeny of animals and plants. Phylogeny is a term coined by Haeckel (1866) to refer to the history of the palaeontological development of species through time and it has been a truism to suggest that fossils are the key to understanding relationships amongst organisms. Nowadays 'phylogeny' has come to have a wider and more varied meaning, simply as the pattern of descent of species whether Recent or fossil. But the role that fossils play in reconstructing phylogeny has remained important to many people and indispensable to a few. The phylogeny of mammals, as expressed through the traditional classifications, is based on the study of fossils (Simpson 1945, 1961). In other groups, where the fossil record is scanty or absent, it has been claimed that phylogeny is unknowable (Newell 1959). The rationale for these viewpoints is found in evolutionary theory and in the belief that fossils are the only direct historical evidence available. Flowing from this is the idea that if evolutionary relationships of ancestry and descent are to be discovered then fossils must be considered important because only they are likely to be ancestors of later fossils and of modern species.

The cladistic method of systematics has challenged this traditional view of fossils and time by drawing a clear distinction between cladograms and trees (Cracraft 1979; Platnick 1979; Fig. 8.1). Cladograms are statements about the distributions of shared characters (see also Chapter 1) and these may be represented as Venn diagrams without implication of evolution. Trees may have the same geometry as cladograms but they carry additional assumptions of historical

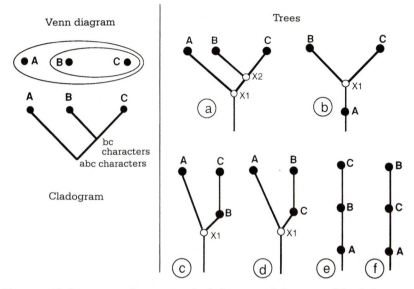

Fig. 8.1 Cladograms and trees. On the left is one of three possible cladograms for three taxa. This may be drawn as a branching diagram or as a Venn diagram of a set and subset. On the right are seven equivalent trees with time as the vertical axis and real or hypothetical ancestors. Tree a, with hypothetical ancestors X1 and X2 is equivalent to the cladogram.

descent with modification. Trees can only be drawn as branching diagrams where the nodes represent ancestors (real or hypothetical), the branching represents speciation (cladogenesis) and the lines represent descent with modification (patristic relationships). There may be several trees which explain the character distribution implied in one cladogram, and the alternatives depend on whether we accept none, one, or more taxa as being ancestral to another.

In traditional evolutionary systematics trees, such as those in Fig. 8.1b–f which include fossils as real ancestors are perfectly acceptable. In cladistic analysis only the tree (Fig. 8.1a) with the same shape as the cladogram is acceptable. This is the phylogenetic tree and is the only one of several alternatives that might be justified by character distribution. For Hennig this was the aim of cladistic analysis. He regarded and explained cladograms as trees, with assumptions about time and evolutionary theory.

An inference about an ancestral species cannot be made from characters alone. Thus, from Fig. 8.1, tree b shows species X_1 as the ancestor of species B and species C. X_1 would have the characters of the group ABC (see cladogram) but lack the characters of B and of C. That is, it would not have any unique characters. Ancestor X_1 would be paraphyletic (see also Chapter 1) and only capable of being recognized with reference to B and C.

An example of such an ancestor might be *Archaeopteryx*. It shares synapomorphies such as feathers and a wishbone with all birds but it lacks the synapomorphies of any subgroup of birds (B and C in Fig. 8.1). Cladistic analysis seeks to discover monophyletic groups. However, ancestral species can never be recognized as such. For this reason putative ancestors − and this usually means fossil organisms − are treated as if they were terminal taxa: that is, as potential sister groups of other taxa. This is suggested even if there are no demonstrable autapomorphies.

8.3 AGE AND RANK

Cladistic classification proposes that sister groups be given equal rank − whether this be Linnaean rank or some numeric rank system (for example Hennig 1969, 1983; Løvtrup 1977). Hennig (1966) argued that this would prove difficult when including fossils in a combined Recent/fossil classification. For example, if we identified a single Carboniferous cockroach species as being the sister-taxon to all modern cockroaches, which are placed in the Order Blattaria, then the single Carboniferous species would be formally classified as an order, leaving many empty ranks (Family, Tribe, etc.). Hennig was also concerned about the equivalence of ranks across all animal and plant groups. For instance, assuming that there is a single tree of life, how then might we compare a speciose extinct group (for example Trilobita) with a species depleted modern group (for example Onychophora)?

Hennig's solution was to suggest that the geological column be divided into six successive time bands and each band be assigned a rank − phylum, class, etc. Groups which had the first fossil representative within any time band would carry a certain rank. Crowson (1970) advocated a very similar scheme but went further by suggesting that this only partly avoided the problem and that, probably, fossils should not be classified with Recent organisms but that separate classifications should be adopted for organisms occurring in each of his time bands. There are considerable problems with both of these solutions (Patterson and Rosen 1977), not the least of which is to deal with the many taxa which span more than one time band.

8.4 STEM GROUPS AND CROWN GROUPS

Hennig (1966) recognized that in the phylogenetic history of any group of modern organisms there is an age of origin of the group and an age of differentiation of the modern members (Fig. 8.2a). The age of origin is specified by the cladogenetic event leading to the modern group we are interested in (Hennig called this the *group) and its modern sister group. The age of differentiation is the age of the latest common ancestor of the modern members of the group. Both of these ages might be inferred by using fossils to give a minimum age in

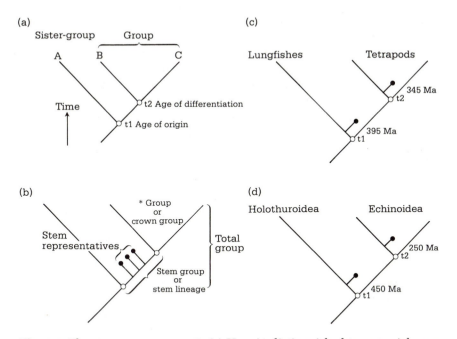

Fig. 8.2 The stem group concept. (a) Hennig distinguished two crucial ages in the development of a Recent group – the age of origin and the age of differentiation. (b) The difference between the age of origin and the age of differentiation may be occupied by fossils. The various names given to segments of history are shown here – see text. (c), (d) Fossils can determine the minimum ages of differentiation and origin leading to the recognition of the length of time occupied by the stem lineage. Two examples are given here (see also Fig. 8.3).

both cases. Two examples are given in Fig. 8.2c,d. Another way in which these ages might be inferred is by using sequence analysis and assuming a molecular clock (see also Chapter 7); but in all probability this clock would have already been calibrated using fossils.

The time gap between age of origin and age of differentiation would have been occupied by fossil species. Hennig referred these fossil species filling this position to the stem-group (Hennig 1969) which he distinguished from the *group. Jefferies (1979) uses the terms stem-group and crown group, the two combined being the total group. All authors have recognized that the stem-group is typological and that, in reality, it would be paraphyletic and consist of a series of fossil species or species groups successively more closely related to the crown group. For this reason Ax (1987) recommends abandonment of the stem-group and replacing it with stem lineage. Individual members would be stem representatives. These terms are illustrated against a tree in Fig. 8.2b.

To include fossil species within the stem lineage requires that at least one synapomorphy is shared with the crown group. To do this the characters of the crown group must already be known. This means that the classification of Recent organisms takes precedence over fossils. Once this is established and fossils are treated as terminal taxa there is no reason why cladistic analysis should not proceed to insert fossils and fossil groups in sequence within the stem lineage (but see below). Naturally, some fossils will be members of the crown group: that is they will belong to the stem lineage of one or more of the contained subgroups. As in other aspects of cladistic analysis it can be seen that the concept of stem representatives and crown group is relative to a particular problem. For example, if the origin of penguins was the problem, then fossil penguins belonging to the stem lineage of modern penguins would be important. But if the problem was the origin of birds then these stem representatives of penguins would be of peripheral interest.

So far, what advantages would be gained by including fossils in stem lineage series? There are three immediate advantages:

1. Fossils may give us minimum ages for cladogenetic events and these, in turn, may be used to calibrate molecular clocks.

2. If stem representatives are arranged in cladogenetic sequence according to synapomorphy then deductions may be made about the order in which modern groups got their characters.

3. This, in turn, allows speculation about character transformations and evolutionary scenarios.

As an example we may look at one of the current theories relating to the origin of the most obvious of tetrapod characters − arms and legs. A deduction from a phylogenetic tree, which includes several stem representative tetrapods, suggests that a fully fledged four-limb condition developed by modification of the pelvic fin to a leg and foot which was followed later by modification of the pectoral fin to an arm and hand. This suggests that legs came before arms and that legs were not necessarily an adaptation to life on land. The animals which bore the first legs may have been primarily aquatic.

It is likely that the longer the time span between the age of origin and the age of differentiation then:

(1) more stem representatives will be found;

(2) more character transformations will take place.

Therefore fossils will potentially have more to offer. In Fig. 8.3 there is an example taken from Smith (1984) of the successive stages in the development of the complex Artistotle Lantern apparatus of echinoids during the very long duration (200 million years) of the stem lineage. Without these fossils it is doubtful if the detailed transformation of test plates to teeth could ever have been worked out.

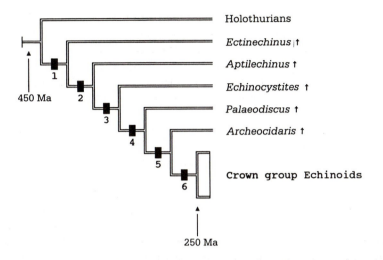

Fig. 8.3 One example of correctly sequencing fossils within a stem lineage to establish the sequence of development of complex crown group characters. The feeding apparatus of modern sea urchins is a highly complex system of modified plates and is called the Aristotle's Lantern. The sequence of fossils shows that this complexity was achieved through a series of sequential modifications. Knowledge of the sequence is important as it allows functional inferences to be made. (1) Lantern consists of hemipyramids; teeth – simple, developed as spines bound together; function scooping. (2) Epiphysis, rotulae. (3) Teeth formed of v-shaped arrangement of rods. (4) Compasses.(5) Curved secondary plates. (6) Tooth plates bound together; Lantern consists of hemipyramids, epiphysis, rotulae; compasses; teeth – flat primary plates, curved secondary plates, function chewing. Example kindly supplied by Andrew Smith and based on Smith (1984).

8.5 INFLUENCE OF FOSSILS ON CLASSIFICATIONS OF RECENT ORGANISMS

With few exceptions most cladists wish to incorporate fossils with Recent organisms in a combined classification but there is a diversity of opinion on what influence fossils may have on classification of Recent organisms. One viewpoint holds that fossils can be interpolated to a classification of Recent organisms but it is doubtful if they can overturn any theory of relationships amongst the Recent taxa (Patterson 1981*a*, Ax 1987). Another viewpoint suggests that cladistic analysis should contain fossils and Recent organisms from the outset and that the inclusion of fossils can overturn a theory of relationship based on the living taxa only. Both viewpoints acknowledge that the inclusion of fossils can influence ideas on polarity of characters and homology.

The first viewpoint is based on the starting point that 'fossils are data in need of interpretation' (Nelson 1978). Fossils are fragmentary remains of hard parts or, at best, incompletely preserved soft parts and these need to be interpreted in the light of a Recent model which is potentially complete with soft parts, molecules and an ontogenetic sequence. The choice of a model can determine the interpretation of the fossil (Rosen *et al.* 1981). The only valid reason for choosing one Recent model rather than another is to recognize a character in the fossil shared with a modern group. And this inevitably means prior recognition of the characters of Recent groups.

Another reason why fossils hold limited information is that their incompleteness means that their relationships can usually only be specified within nodes. For example, it may be possible to identify a synapomorphy of a fossil and a crown group but it may not be possible to check if the relevant synapomorphies of the various crown subgroups are truly absent or simply uncheckable. Thus, fossils may give less precision to any analysis than Recent organisms.

The second point of view advocates that fossils can overturn theories of relationships of Recent organisms. This view has arisen with the development of numerical cladistics. Gauthier *et al.* (1988) and Doyle and Donoghue (1987) have argued that, as units of analysis, fossils are no different from Recent organisms and that their inclusion in a combined analysis is preferable to separating the two.

Two examples have been used to support this viewpoint and to provide discussion about the role of fossils in numerical cladistics − amniotes and seed plants. Gauthier *et al.* (1988) produced a cladistic classification of five Recent groups of amniotes using 109 characters, 45 of which were skeletal and potentially capable of fossilization. The remainder were characters of soft anatomy. The resulting tree is shown in Fig. 8.4a where birds and crocodiles are sister groups and mammals are the sister group to those two combined. They then added to the data matrix 24 extinct taxa which through traditional classifications are considered stem representatives of crocodiles and of mammals. The data matrix was enlarged by the addition of skeletal characters to contain 274 characters. With this new matrix including the fossils, the relationships among the Recent groups changed so that mammals became the sister group to the remainder of the amniotes (Fig. 8.4b).

Through additional series of analyses involving selective addition of the fossil taxa they found that certain fossils were pivotal; inclusion meant the tree topology went one way, exclusion meant that it went the other. They found that within the long series of stem lineage representatives those near the base or those near the crown group had no effect in changing topology but that any one of the middle series (shaded in Fig. 8.4) resulted in the change in tree topology among the Recent taxa. Certain observations on the study of Gauther *et al.* are worth noting. The increase in the number of taxa increased homoplasy. This, in itself, is hardly surprising: the phenomenon has been well documented (Farris 1972; Sanderson and Donoghue 1989) and will happen whether the added taxa are fossils or not. But the addition of fossils changed the optimization of characters, such that minimization of character changes (reversals and convergencies)

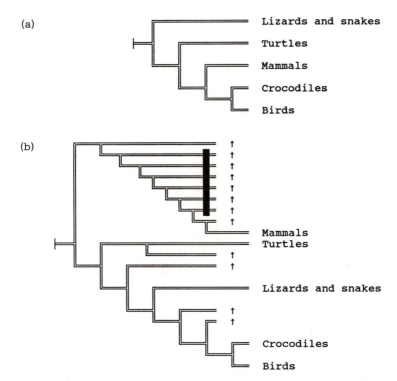

Fig. 8.4 Amniotes and the influence of fossils on topology. (a) Gauthier *et al.* (1988) parsimony analysis of five Recent taxa (109 characters). (b) The same five taxa with the addition of extinct taxa and 274 characters resulted in this tree. The addition of any of the taxa embraced with the bar caused the topology amongst Recent taxa to switch from that in (a) to that in (b). Simplified from Gauthier *et al.* (1988).

required a change in topology. A theoretical example is presented in Fig. 8.5. Many of the added fossils in the amniote example were prone to do this because they had new and very different character combinations. More usually than not they were highly plesiomorphic for many of the character states and derived in a few. Many fossils are understandably of this kind and this may be their uniqueness.

The second example designed to 'test' the influence of fossils in numerical cladistic analysis is that of the relationships amongst seed plants (Doyle and Donoghue 1986, 1987; Donoghue *et al.* 1989). This example is more complicated because there were no single most parsimonious trees resulting from any of the analyses, such that addition of fossils did not arbitrate between alternate groupings. But Doyle and Donoghue suggested that inclusion of fossils did arbitrate between equally parsimonious theories of character evolution. They cited the example of leaf evolution (Fig. 8.6).

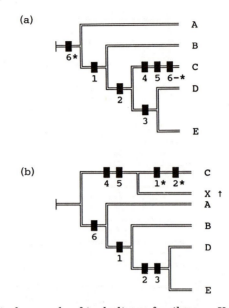

Fig. 8.5 Theoretical example of including a fossil taxon X having a character combination which alters the topology amongst Recent taxa A–E (a) This is the most parsimonious solution involving a reversal in character 6. (b) The fossil taxon X is apomorphic for characters 4 and 5 but otherwise plesiomorphic. Taken from Donoghue *et al.* (1989).

These 'experiments' with adding/deleting fossil taxa in one sense, are hardly a test of the potential of fossils to change topology because at each stage of addition or deletion we are changing the nature of the problem. The amniote example changes from a 5-taxon problem to a 29-taxon problem and the seed-plant example changes from a 7-taxon problem to a 20-taxon problem. From these 'experiments' it appears that the potential for fossils to change or clarify our ideas of character evolution is more important than the potential to overturn theories of relationships between Recent groups.

8.5.1 Missing data

Fossils can influence numerical cladistic analyses through character coding. Fossils may be very incomplete and therefore may have many '?' codings for many of the characters (soft anatomical, some skeletal/hard part anatomy, molecular). Introduction of taxa with many '?' codings creates problems of resolution.

While incompleteness has been recognized by most workers others have argued that incompleteness may instead be a feature of Recent animals in an analysis which includes both fossil and Recent organisms. Gauthier *et al.* (1988)

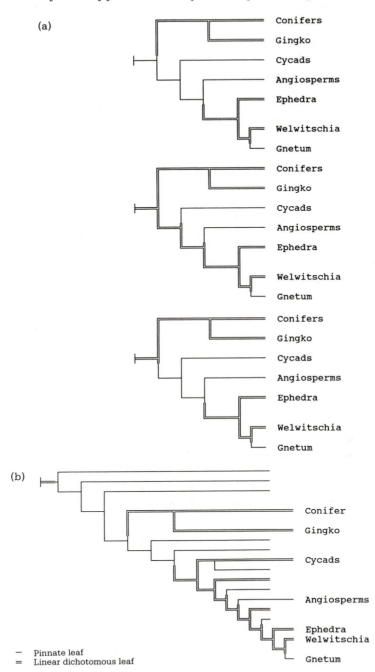

Fig. 8.6 Introduction of fossils may help to arbitrate between equally parsimonious optimizations. (a) Three different ways to optimize the character pinnate leaf. (b) The introduction of fossils determines that the pinnate leaf is primitive for seed plants. Example from Donoghue et al. (1989).

pointed out that morphological divergence between stem representatives and modern species may mean that many features have been lost or modified in such a way that they cannot readily be compared between fossil and Recent representatives. For example, if we were interested in the relationships of modern lampreys and hagfishes to one another and to modern jawed vertebrates then we could use comparable details of soft anatomy – muscles, nerves, molecules, etc. But when we introduce fossil forms, many of which are Palaeozoic armoured fishes, outwardly dissimilar to naked lampreys and hagfishes, then there is very little basis for comparison. In a data matrix combining fossils and Recent organisms many of the details about the complicated armour would be entered as 'not applicable' against the Recent lampreys and hagfishes which are without armour. In this sense, lampreys and hagfishes carry a high burden of 'missing' information. For computer algorithms (for example PAUP and Hennig86) 'missing' and 'not applicable' are treated the same way; such codings allow the algorithm to choose alternative states (but see Platnick *et al*. 1991). In the amniote example of Gauthier *et al*. they found that, within their combined data matrix, Recent mammals and turtles were each 'missing' 15 per cent of the data. The 'missing' data was due entirely to non-applicable scorings. The missing data in fossils was due to both 'non-applicable' and non-preservation. It might be possible that, given a matrix of only hard parts, the fossil taxa may have 100 per cent data and Recent taxa may be considered more incomplete. However, this will only happen when all of the organisms are treated as 'fossils'; that is, as organisms without soft parts or molecules.

8.5.2 Basal taxa

From the discussion above it has been pointed out that plesiomorphic stem representatives can influence polarity decisions, theories of homology and, perhaps, topology amongst trees containing Recent organisms. They are therefore seen to be very important by some workers and necessary inclusions in any cladistic analysis. This has been formalized by using the term 'basal taxa' (Cloutier 1991), defined as that taxon (species or higher group) which is separated from the stem species by one cladogenetic event: or in terms of characters, 'as that taxon which shares the synapomorphies which diagnose the clade but which lacks the synapomorphies derived within the in-group' (Trueb and Cloutier 1991). Phrased in terms of characters then basal taxa are, by definition, paraphyletic (see also ancestors above). Basal taxa are the first 'in-group' taxa and may have importance in changing our ideas of homology (Maddison *et al*. 1984). For example, Rosen *et al*. (1981), in their study of the origin of tetrapods suggested that one of the synapomorphies of Recent lungfishes and tetrapods was the fusion of the palate with the neurocranium. Subsequent to that study, a basal taxon lungfish (*Diabolepis*) was discovered in Devonian rocks of China and shown to have a palate free from the neurocranium, just like most other fishes. The theory that 'fused palate' was a synapomorphy of lungfishes + tetrapods was falsified. Because of this, some workers argue that, when analysing the relationships between groups then the coding for those groups, such as lungfishes or tetrapods, should be based on the basal taxa (Schultze 1987). This approach may have some evolutionary appeal in trying to 'catch the lineage in its early phases of differentiation' but it begs the question as to how the basal taxa are to be recognized without a prior tree or theory of relationship in mind. That prior tree comes from some theory of relationship amongst the crown groups.

8.5.3 Fossils and stratigraphy

Throughout this chapter little has been said directly about fossils and stratigraphy. Hennig thought that the stratigraphic position of fossils may be important information when trying to polarize character states. It is impossible to read a phylogenetic sequence directly from the rocks using the stratigraphic record alone. Some morphologically-based theory of relationship is needed, to be certain that younger brachiopods are not being compared with older molluscs. However, assuming that the morphological groups are roughly delimited, then the idea of what comes earlier is necessarily more plesiomorphic has long been a tenet of palaeontology (Bretsky 1975). This was phrased in cladistic terminology by de Jong (1980) as: 'If in a monophyletic group a particular character condition occurs only in older fossils and another condition of the same character only in younger fossils, then the former is considered the plesiomorphous and the latter the apomorphous condition of that character'. We could substitute 'Recent' for 'younger fossil' because many palaeontologists believe that fossils are plesiomorphous when compared with their Recent counterparts.

The generality of this belief is not in doubt, but the applicability in any one instance is always fraught with difficulty. If we accept this tenet then we assume that the sequence of fossils we recover from the rocks is a true historical sequence and doubts about the incompleteness of the fossil record must be put aside.

Some workers (for example Hill and Camus 1986) have argued that finding fossils out of expected stratigraphic sequence might be used to test a theory of character evolution or even to test theories of the cladistic sequence of stem representatives. Again, we must assume that the fossil record is capable of yielding an accurate record.

As a compromise, Fortey and Jefferies (1982) suggested that stratigraphic occurrence may help to 'fine tune' a cladistic analysis based on morphology alone. The stratigraphic sequence of fossils, they argue, may be used to establish species lineages when the assumed morphological changes are small (the example they choose is the reappearance of genal spines in derived members of a trilobite lineage). Again, they stress that this approach may only be valid when the stratigraphic record is demonstrably good and the sequence repeatable from place to place.

For this reason there have been several attempts to provide estimates of the completeness or otherwise of the fossil record (Paul 1982; Hay 1972) or the quality of the record (see discussion in Schoch 1986). It might therefore be possible to express the stratigraphic test in some probabilistic formula. At present the stratigraphic criterion has few adherents. The general consensus is that using the stratigraphic record to establish polarity or to test theories of relationship is always suspect, since any false or unwanted results may be explained away by invoking incompleteness of the record. It is possible that with more precision stratigraphic criteria may have a part to play in cladistic analysis.

8.6 CONCLUSIONS

Fossils should be included in combined Recent–fossil classifications. To do this they should be treated as terminal taxa and coded with reference to Recent taxa. They will inevitably have missing values, particularly if combined with molecular data.

The advantages of including fossils are: estimations of sequences of character acquisition and of character transformations and minimum ages for cladogenetic events; these might be useful in calibrating molecular clocks and dating events in biogeographic history; evaluation of theories of homology and polarity of character-state changes.

9.
Cladistic biogeography

Christopher J. Humphries

9.1 INTRODUCTION — LIFE AND EARTH TOGETHER

Historical biogeography has always been dependent on systematics. It has long been recognized that the distributions of organisms and areas of endemism are not random but show distinct and highly repetitive patterns which require an historical explanation. The developments in panbiogeography (Croizat 1964) and cladistic biogeography (Nelson and Platnick 1981; Humphries and Parenti 1986) have led to hypotheses that the Earth and its biota have evolved together and share a common history. Consequently, the methods of systematics can be applied to biogeography and similar problems of co-evolution which require the comparison of two or more groups of organisms considered to share similar histories. The question to ask now is how can biogeographic theories be erected from biotic data. Do two or more cladograms say anything about area relationships?

The developments in the theory of continental drift and its general acceptance during the 1960s led to the acceptance of the idea that disjunct biotic patterns and corresponding geological patterns were due to the same causes of Earth history. The idea of the Earth constantly changing means that many, if not all, of the migration solutions to biogeographic problems are wrong. A possible solution emerged when Brundin (1966) applied Hennig's (1950) definition of relationship to the problems of vicariant distribution of southern hemisphere chironomid midges. The ensuing developments in theory fused the track approach of Croizat's panbiogeography with cladistics to eventually produce cladistic biogeography. For cladistic biogeographers systematic patterns become understandable and comparable when they are expressed as area cladograms. Area cladograms, branching diagrams of areas, express the interrelationships of areas as determined from systematic information by substituting the taxa for the areas in which they occur.

It will be shown here that several different taxonomic groups often demonstrate the same pattern of area interrelationships and it is possible to make a general hypothesis about the interrelationships of biotas and areas of endemism. Corroborated hypotheses of this sort can be compared with similarly organized, but independent, information from geology.

9.2 CLADISTICS AND BIOGEOGRAPHY

The earliest coherent applications of cladistics to biogeography were by Hennig (1966) who used cladograms to determine the 'centres of origin' of monophyletic groups. Hennig (1966) reasoned that there is a close relationship between species and the space that each one of them occupies. But rather than saying that species and spaces evolve together, Hennig continued the nineteenth century idea that dispersal patterns are unique for each taxonomic group and each has an independent history.

The best applications of Hennig's method are found in Brundin (1966, 1972*a,b*, 1981) for chironomid midges and Ross (1974) for caddis flies. Although Hennig, Brundin, and Ross brought much greater precision to biogeography by superimposing areas onto phylogenies and inferring the lowest number of dispersals for each group, their method relied on *ad hoc* assumptions that groups have 'centres of origin' and species migrate from the centre to distal areas.

9.2.1 The progression rule

Central to Hennig's (1966) method to find a 'centre of origin' for a group of taxa from a particular cladogram, was the idea that phylogenetically primitive members of a monophyletic group will, by definition, be found near that centre. In other words, within a continuous range of different species of a monophyletic group it was considered possible in certain circumstances that a transformation series of characters would run parallel with progression in space, such that the youngest members would be on the geographical periphery of a group. A good example is given by Ross (1950) with the *Wormaldia kisoensis* complex of caddis flies. The geographical distribution of the nine species, eight species in the Asian Pacific between Sarawak in the south and Japan in the north and one species in the Smoky mountains of eastern North America, is shown in Fig. 9.1. The cladogram of phylogenetic relationships is shown in Fig. 9.2. Ross (1974) assumed that the base of the stem denotes the ancestor of the whole group. That the most derived species pair occur in Japan and eastern USA led Ross to consider the simplest dispersal hypothesis of an origin for the group in Asia and a single dispersal event for one species across the Bering Strait to North America.

Brundin's classic studies of chironomid midges (Brundin 1966, 1972*a,b*, 1981) showed that the southern temperate areas of South America, southern Africa, Tasmania, south-east Australia and New Zealand are inhabited by 600 to 700 species. Trans-antarctic relationships are a recurring phenomenon throughout the group, so by way of an example consider the midges of subfamily Diamesinae, which display a double extra-tropical distribution: two major groups present in both northern and southern hemisphere temperate areas but absent from the tropics. The largest, and most widespread is the relatively

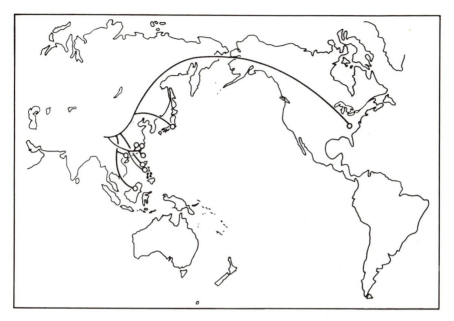

Fig. 9.1 Distribution and phylogeny of *Wormaldia* (Caddis flies). The circle in Japan is *W. kisoensis*, that in eastern North America, *W. mohri*. (After Ross 1974, p. 217.)

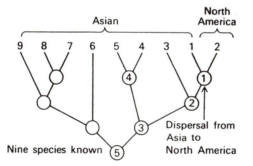

Fig. 9.2 Cladogram for the nine species of *Wormaldia* (see Fig. 9.1). (After Ross 1974, p. 216.)

generalized tribe Heptagymi, represented by 11 species in Andean South America, two species in south-eastern Australia and five species in New Zealand. Its sister group is the relatively more apomorphic and monotypic tribe Lobodiamesini of New Zealand.

The cladogram in Fig. 9.3 shows that there are a total of 25 terminal taxa in the southern hemisphere areas of South America, New Zealand, Australia and

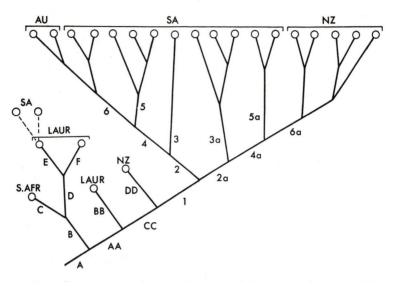

Fig. 9.3 Partial reconstruction and area cladogram of the subfamily Diamesinae (Diptera, Chironomideae). (After Brundin 1981, fig. 3.7, p. 119.)

South Africa, and three groups in Laurasia. The monotypic genus *Heptagyia* occurs in South America and *Paraheptagyia* has five South American species. According to Brundin the south-eastern Australia subgroup of two species is a younger evolved offshoot of the older South American group including *Heptagyia*. Brundin considers the Australian taxa to have dispersed from Patagonia or east Antarctica at stem 6 by the end of the Palaeocene because they have derived characters and because the stem species (indicated by 1, 2, and 4) never occurred in Australia. The other stem (2a) includes *Reissia* with three species in South America, *Limaya* with two species in South America, and *Maoridiamesa* with five species in New Zealand. Brundin (1981) considers the fact that *Maoridiamesa* is on a different stem from the Australian *Paraheptagyia* group agrees well with plate tectonic theory for an early separation of New Zealand from western Antarctica in the Upper Cretaceous. The fact that *Maoridiamesa* is a comparatively younger, derived offshoot of an older group in South America is, according to Brundin, evidence of long distance dispersal from South America via west Antarctica to New Zealand of stem species 4a rather than a vicariance event. In other words, Brundin thinks that the *Maoridiamesa* group is younger than the areas in which it occurs.

The interpretation of cladograms as phylogenetic trees rather than as hierarchical groups with relatively more inclusive groups of set membership, often requires *ad hoc* assumptions not fully justified by the information on which they are based to be made. Furthermore, the interpretation of individual cladograms as having individual histories leads to certain conceptual difficulties. A crucial one is the repetition of distribution patterns. If we have, as is this case, many unrelated taxonomic groups repeating a pattern of distribution between major continents, such as South America and Australia and New Zealand, it is improbable that there were many dispersal events, with each group separately making its way from one continent to the other. The most logical and simplest conclusion would be to suggest that at one time, the continents were in contact and that the present day pattern was due to the break up of a formerly continuous biota.

9.2.2 Vicariance biogeography

The breakthrough in the application of cladistic reasoning to biogeography came with the efforts of biogeographers such as Nelson and Rosen in their interpretation of Croizat. Instead of a 'vacuum' theory of biogeography, whereby certain areas were originally devoid of taxa later to be colonized from other source areas, there was proposed an equally plausible alternative. Disjunct distributions could come about by vicariance events because their ancestors orginally occurred in the areas where they occur today, and the taxa evolved in place (Croizat *et al.* 1974). In other words, dispersal models explain disjunctions by dispersal across pre-existing barriers, vicariance models explain them by the appearance of barriers fragmenting ancestral species ranges. So what became particularly clear was the important idea that distributional data are insufficient to resolve decisively either dispersal or vicariance as the cause of a disjunct distribution pattern. Therefore, when faced with a particular distribution, as Platnick and Nelson (1978) argued, one should not worry about its cause but whether or not it conforms to a general pattern of relationships shown by other groups of taxa endemic to the areas occupied. Thus, as in cladistics where three-taxon statements are the most basic units for expressing relationships, in biogeography, three-area statements are the most basic units for expressing area relationships.

Generality of the area cladograms can be examined by comparison with other unrelated taxonomic groups endemic to the relevant areas and corroboration of a particular pattern is equivalent to a general statement for the relative recent ancestry of the biotas under examination.

Initial applications of the method (for example Rosen 1976) encountered problems of incongruence and unresolved statements in the general area cladograms. Theoretically, it should be possible to connect every area of endemism into one larger general statement of interrelationships. However, our perception of the world is less than perfect for a variety of reasons — extinction, dispersal of widespread taxa, and restricted distributions of taxonomic groups.

9.2.2.1 Rosen's method (1976)

Rosen set out to apply Croizat's (1952, 1964) vicariance method to Caribbean biogeography. The application has special significance because it was the first concise exposition of panbiogeography with cladistics added, so that the groups considered are monophyletic rather than both monophyletic and paraphyletic, as in many of the Croizat's former examples. The method consisted of marking the distribution of disjunct components of monophyletic groups on a map, and linking the areas of each group by a line or track. Where tracks linking sister taxa (fossil or Recent) coincided repeatedly, the lines delimit a generalized track, assumed to link two or more biotas that are vicariant products or fragments of a single ancestral biota. The sequence of events was compared to the data from historical geology to account for the fragmented pattern.

Rosen recognized four generalized tracks for the Caribbean region: a North American Caribbean track and a South American Caribbean track, both mainly terrestrial; and an eastern Pacific–Caribbean track and an eastern Atlantic–Caribbean track, both mainly marine (Fig. 9.4). His sources of distributional data include plants, amphibians, reptiles, birds, mammals, and fish.

The main problem was to determine what these tracks meant in terms of distributional history. Rosen suggested that the four generalized tracks may be

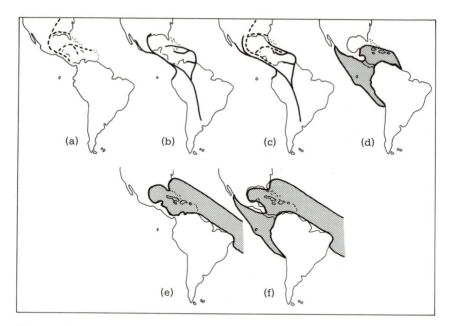

Fig. 9.4 Summary of transcontinental generalized tracks. (a) North American – Caribbean track. (b) South American – Caribbean track. (c) Overlapping of North and South American tracks enclosing the Caribbean sea. (e) Eastern Pacific – Caribbean track, the trans-oceanic track hypothesized to be the youngest here. (d) Eastern Atlantic – Western Atlantic track, a track of intermediate age. (f) Eastern Pacific – eastern Atlantic track, hypothesized to be the oldest trans-oceanic track depicted here. (After Rosen 1976, fig. 6, p. 444.)

interpreted either as a result of four separate dispersal routes into the Caribbean areas or as vicariance events that subdivided ancestral, marine, and terrestrial biotas into smaller units. The first interpretation demands that dispersal was remarkably co-ordinated for such vastly different groups as birds, plants and amphibians. The vicariance interpretation required geological events that isolated the eastern and western Atlantic, and the eastern Pacific from the Caribbean followed by an intermingling of North and South American biotas in the Mesoamerican region. Rosen (1976) regarded dispersal hypotheses as untestable because they appeal to individual explanations.

There is another problem, however. Tracks cannot be refuted in the same way as cladograms since tracks only connect sets of areas diagnosed by monophyletic groups. Tracks are not meaningful ways of expressing relationship in the same sense as cladistics. It appears that generalized tracks have been considered to be many things, such as statistical measures of similarity between disjunct biotas (Ball 1976), as measures of overall similarity as in phenetics (Patterson 1981*b*), or minimum spanning trees (Page 1987; Henderson 1989). Since tracks are defined on the connections of related taxa between disjunct areas they do to some extent represent a similar generalization in geography as cladograms do in systematics. Tracks as used by Croizat (1952, 1958, 1964), although usually monophyletic, are not necessarily so. Since they are not hierarchical, but maybe minimal spanning graphs connecting nearest neighbours, they show a more general resolution of geographical patterns than cladograms.

To improve resolution it seems that the same type of information is required in biogeography as used in cladistics. To make satisfactory comparisons between organisms and areas it is necessary to have cladograms of areas that can be derived from the taxonomic cladograms − a technique less generalized than track analysis.

9.2.2.2 Cladistic biogeography − Platnick and Nelson (1978)

Nelson and Platnick proposed a solution for a definition of area relationship by asking the question: 'Why are taxa distributed where they are today?' They gave two possible answers: either they evolved *in situ* or they evolved elsewhere and dispersed into new areas. The difference between vicariance and dispersal lies in the relationship between the age of a taxon and the age of the barriers limiting the area. Vicariance predicts that taxa in two (or more) areas and the barriers between them are the same age; whereas dispersal always predicts that the barrier predates the taxa.

Their method involved finding monophyletic groups with taxa occurring in at least three or more similar areas. Cladograms are produced for each group of organisms. The names of the taxa at the terminal tips of the cladograms are replaced by the names of the areas in which each taxon occurs. These are area cladograms, and the tips of the branches thus become areas as diagnosed by the taxa. The sum of the areas on one cladogram is equivalent to a track. The sums of similar areas on several cladograms are equivalent to generalized tracks. To obtain a cladogram of biotas the individual cladograms are combined. Consider three areas of endemism, A, B, and C, in which occur a group of freshwater fishes F1−3 and a group of trees T1−3 (Fig. 9.5a,b). The characters of each group are analysed and cladograms produced (Fig. 9.5a,b). The area of each taxon is then substituted onto the cladogram. The area cladograms are then compared. In this case they are congruent (Fig. 9.5c). The hypothesis for the three areas in this hypothetical example is that B and C share a more recent history than either do with A for these two groups.

The success of finding a congruent vicariant pattern in nature depends on the frequency with which common factors have affected the phylogeny and distribution of two or more groups of organisms. To find congruent patterns Rosen (1978) considered it necessary to delete unique, unresolved, or incongruent components from cladograms. By way of an example, consider the following five areas of endemism, A−E, and the two monophyletic groups of fishes (F1−4) and trees (T1−4; Fig. 9.6a,b). The area cladograms (Fig. 9.6a,b) are not completely congruent since the fishes and trees each have an unique species in areas A and E respectively. Rosen's approach was to delete the incongruent portions of the cladograms to make a reduced area consensus tree (Fig. 9.6c).

9.2.2.3 Poeciliid fish in Middle America − Rosen's example (1978)

To undertake cladistic biogeography at least two groups of taxa must be available for the same or similar set of areas. Rosen's analysis of poeciliid fish in Middle America was the first practical example of applying the Platnick and Nelson (1978) method to real organisms.

Rosen (1978, 1979) examined *Heterandria* and *Xiphophorus*. Both genera have close relatives elsewhere but each has a monophyletic sub-group inhabiting the same 11 general areas in southern tropical Mexico, south to eastern

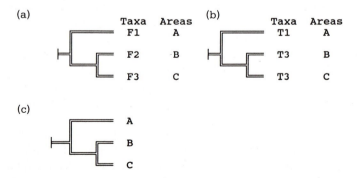

Fig. 9.5 Cladograms and area cladograms for two sympatric groups occurring in three areas of endemism, A, B and C. (a) Fishes. (b) Trees. (c) Nelson consensus tree (see also section 9.2.2) derived by combining the components of area cladograms in (a) and (b).

Honduras in *Xiphophorus* and further south to eastern Nicaragua in *Heterandria* (Fig. 9.7). Cladograms for the two genera are shown in Fig. 9.8a,b. The cladograms for both genera are converted into area cladograms (Fig. 9.8c,d). Next, simplified area cladograms were produced allowing only one term for each area

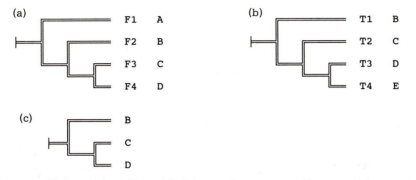

Fig. 9.6 Cladograms and area cladograms for two partially sympatric groups occurring in a total of five areas, A–E. (a) Fishes. (b) Trees. (c) Reduced area Nelson consensus tree (see also section 9.2.2) derived from combining the area cladograms.

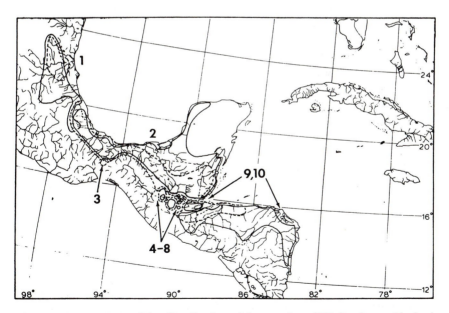

Fig. 9.7 Comparison of the distribution of the species of *Xiphophorus* (dashed and *Heterandria* (solid) in Middle America. Numbers refer to areas defined by the occurrence of taxa. See text for explanation. (After Rosen 1979, fig. 45, p. 367.)

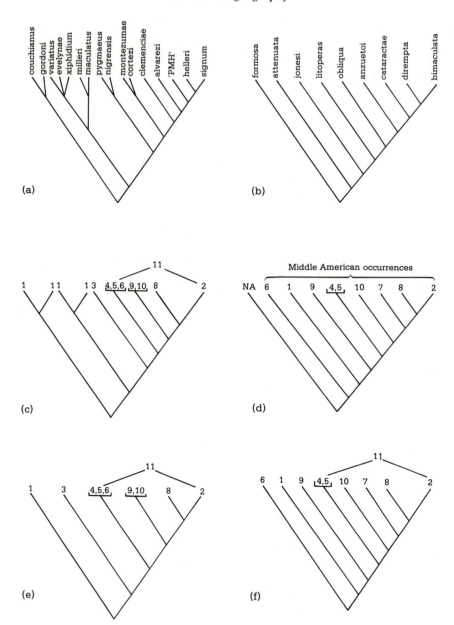

Fig. 9.8 Rosen's method. Species cladograms for *Xiphophorus* (a) and *Heterandria* (b) (After Rosen 1979, figs 48 and 49, pp. 371–2). Area cladograms for *Xiphophorus* (c) and *Heterandria* (d) (After Rosen 1979, figs 48 and 49, pp. 371–2). Simplified area cladograms for *Xiphophorus* (e) and *Heterandria* (f) removing redundant areas. (After Rosen 1979, figs 48 and 49, pp. 371–2.)

(Fig. 9.8e,f). A reduced area cladogram common to both groups was then produced by deleting unique information from each cladogram (Fig. 9.9). According to Rosen, unique information does not contribute to the shared history and the common pattern was then inferred to reflect the history shared by *Xiphophorus* and *Heterandria*.

Rosen (1978, 1979) produced a reduced area cladogram to generate a biogeographical hypothesis for *Xiphophorus* and *Heterandria* because the individual cladograms were incongruent for certain areas, in this case brought about by widespread taxa. Cladistic biogeography would be uncomplicated if all groups of organisms were each represented by one, and only one, taxon in each of the smallest identifiable areas of endemism, but this is not the case. Unique components in individual cladograms occur for a variety of reasons: failure of populations to divide in response to the formation of natural barriers, and dispersal or extinction in one or more areas. The problem with Rosen's reduced area cladograms is that they delete data, a tree pruning method of comparison (see Page 1988). Deleting non-congruent elements of different patterns is akin to throwing away information. Unique patterns may be meaningful and cannot

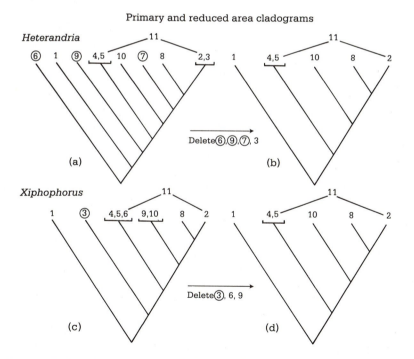

Fig. 9.9 Reduced general area cladogram showing components common to both (a,b) *Heterandria* and (c,d) *Xiphophorus*. (After Rosen 1979, figs 48 and 49, pp. 371 and 372.)

at the same time be incongruent (Platnick 1981). Several solutions to this problem have been suggested (see Mickevich 1982; Wiley 1988; Brooks and McLennan 1991) but only two are described here: Wiley's (1980, 1981) ancestral maps method and component analysis (Nelson and Platnick 1981; Page 1989a).

9.2.2.4 Ancestral species maps – Wiley's method (1980, 1981)

When reviewing Rosen's *Xiphophorus* and *Heterandria* data, Wiley (1980, 1981) considered that all of the speciation events for the Middle American monophyletic groups were successive events. All of the dichotomies in the cladograms are equivalent to the formation of natural barriers followed by speciation in both fish genera. The generalized track of both groups was equivalent to two ancestral taxa. Although this introduced interpretations not necessarily supported by the data Wiley considered the assumption valid since the overall distribution of both groups extended well beyond the Middle American region.

This assumption overcame the problem of unique events with each taxon (Fig. 9.10). The first speciation events were unique for both *Xiphophorus* and *Heterandria*, the origin of *Xiphophorus pygmaeus* and *X. nigrensis* and the ancestor of all other swordtails, from the swordtail common ancestor, and a unique event which isolated *H. attenuata* in area 6 (Fig. 9.10b). The second event was common to both genera and involved the vicariance of the whole of area 1 from the rest of the ancestral range. This resulted in the origin of *H. jonesi* and the ancestor of the two remaining *Xiphophorus* species in area 1 (*X. montezumae* and *X. cortezi*), although Wiley did not include it on his maps. The subsequent division of these two species further subdivided area 1. It should be concluded that the vicariance event dividing *Xiphophorus* into the two areas 1a and 1b did not affect *Heterandria* at all.

At the next step Wiley believed that two separate and unique events occurred in each genus. *Xiphophorus clemenciae* originated in area 3 and *Heterandria litoperas* originated in area 9 (Fig. 9.10d). The next event was common to both groups and involved a separation of a central western part of the ancestral range (Fig. 9.10e). The next event separated the southern portion of the remaining ancestral range and the origin of the *Xiphophorus* 'PMH' and *Heterandria anzuetoi* (Fig. 9.10f). The next event was the unique origin of *H. cataractae* in area 7 (Fig. 9.10g), and finally the peripheral isolation and origin of *Xiphophorus signum* and *Heterandria dirempta* occurred which were isolated in area 8 from the remaining taxa, *Xiphophorus helleri* and *Heterandia bimaculata*, in area 2 (Fig. 9.10h).

To obtain a single hypothesis for the interrelationships of the areas of endemism the cladogram in Fig. 9.11 summarizes the sequence of events for Wiley's successive speciations from an ancestral species. The result provides a complete hypothesis for all 11 areas and is, in this sense, more complete than Rosen's (1978) analysis. However, it must be criticized because the method

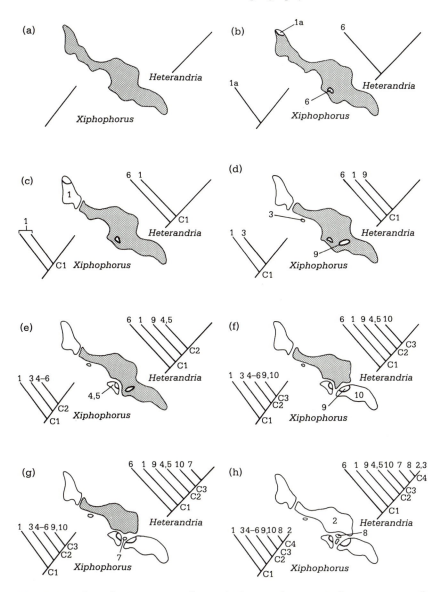

Fig. 9.10 Inferred sequence of speciation and ancestral area maps for *Xiphophorus* and *Heterandria*. In each diagram stippled areas are inferred ancestral ranges. The phylogenetic position of this ancestral species is represented by the unnumbered branch in the area cladograms above and below the geographic map. Numbered branches correspond to the species represented by the number of the areas they inhabit. Common speciation events in the history of the two groups are labelled C1, C2, etc. These labels correspond to the original area cladogram of each species group. (After Wiley 1981, fig. 8.16, p. 303.)

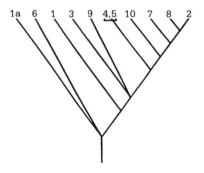

Fig. 9.11 Consensus cladogram for *Xiphophorus* and *Heterandria* based on Wiley's ancestral area maps. See Fig. 9.10 and text for explanation.

introduces evolutionary assumptions and a degree of interpretation not supported by the data.

9.2.2.5 Component analysis

Incongruence between two or more cladograms can occur for a host of historical reasons. Different groups of organisms exhibit a variety of older or younger patterns than the groups to which they are being compared, widespread taxa, redundant information, extinction and unresolved taxonomic groups. Variation in the different patterns, effectively the same as sampling errors, lead to errors in predicting patterns of area interrelationship.

The problems of redundant, missing, and ambiguous information can be assessed with component analysis (Platnick and Nelson 1978). To demonstrate component analysis consider four areas of endemism, A—D and the hypothetical distributions of three different monophyletic groups occurring in those areas of endemism (Fig. 9.12). All three groups, lizards, frogs, and birds, each have an endemic species and two widespread species in two and three areas, respectively (Fig. 9.12). Nelson (1984), who introduced this example, noted that it was as intuitively opaque to analysis as he could possibly devise, since it contains almost all of the difficulties encountered in biogeographic analysis.

Missing areas   The cladograms in Fig. 9.12a—c all have three areas, but area D appears to be 'missing' from the lizard cladogram, and area A is missing from the frog and bird cladograms. As Page (1988) pointed out the problem of 'missing' areas is trivial at one level, in the sense that if an area is not present it cannot provide any information as to area relationships. However, as shown below, the problem is acute when it is necessary to combine two or more cladograms.

Widespread taxa and redundancy   Platnick and Nelson (1978) argued that widespread taxa, as compared with endemics are uninformative, by comparison

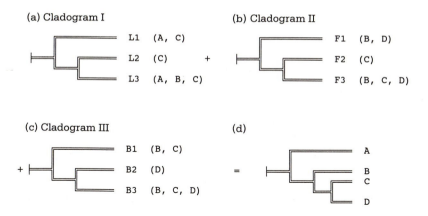

Fig. 9.12 Cladograms and area cladograms for three groups occurring in a total of four areas, A–D. (a) Lizards. (b) Frogs. (c) Birds. (d) Nelson consensus tree (see also section 9.2.2) derived by combining components of cladograms derived under 'assumption 2'.

to area relationships derived from endemic taxa. Widespread taxa introduce complications into biogeographic analysis because they obscure resolution and introduce redundancy in the sense of representing areas more than once on the terminal of any given cladogram. Widespread taxa occur for a variety of historical reasons, as outlined in section 9.2.2.5 above. Platnick and Nelson concluded that areas have to be treated quite differently from taxa, but only taxonomic hypotheses can say anything about the interrelationships of areas. To overcome problems of widespread taxa and redundancy Nelson and Platnick (1981) introduced the rather cryptically entitled concepts, assumptions 1 and 2.

9.2.2.6 Assumption 1

Consider Fig. 9.12a, cladogram I showing the interrelationships of three lizards, L1–3. L1 occurs in two areas, A and C, L2 in area C, and L3 is widespread in areas A–C. Under assumption 1, Nelson and Platnick (1981) argued that whatever was true for L1 in area A in terms of its relationships with L2 and L3 in areas A, B, and C, was also true for L1 in area C. In other words, even if L1 could be resolved into two taxa, say Lx and Ly, each occurring in areas A and C respectively, the general relationships could either be monophyletic or paraphyletic (Fig. 9.13). Similar arguments can be applied to L3 in areas A–C, as summarized in Fig. 9.17b (and the other cladograms, II and III). The three occurrences of C in L1–3 and the two occurrences of A for L1 and L3 could be explained by invoking the extinction of L1 in area B and L2 in area A and C, leaving the one occurrence of C for L2 and the occurrences of A and C for L1 as relics of an older pattern. Consequently, assumption 1 assumes that only speciation and extinction have occurred (Page 1988; Fig. 9.13d).

9.2.2.7 Assumption 2

Consider again the lizards in Fig. 9.12a. Under assumption 2, Nelson and Platnick (1981) argued that whatever was true for L1 in area A in terms of its relationships with L2 and L3 in areas C and A–C respectively, was not necessarily true for L1 in area C. Consequently, considering L1 as having two quite separate start points there are now two possibilities for further biogeographic analysis (Fig. 9.14). This suggests that for L1 either area A or C is redundant in terms of area relationships but at this point we do not know which it is. Unlike assumption 1, assumption 2 accommodates speciation, extinction, dispersal, and failures to vicariate, thus in terms of the original cladogram some of the areas can have had a polyphyletic origin. Considering cladogram I as a whole gives six possible area relationships for further analysis, although in this example

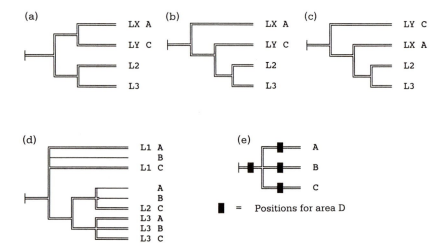

Fig. 9.13 (a–c) Three cladograms derived from the lizard cladogram in Fig. 9.12a under 'assumption 1'. (d) The triple occurrence of area C in L1–3, and the double occurrence of area A in L1 and L3 explained by extinction (= single line) of L1 in area B and L2 in areas A and B. (e) Cladogram available for further biogeographic analysis under assumption 1.

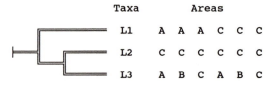

Fig. 9.14 Six possible area cladograms derivable from the lizard cladogram in Fig. 9.12a under 'assumption 2'.

there is only one, (A(C B)) which is a three-area statement and worthy of further analysis (Fig. 9.17d).

9.2.2.8 Assumption 0

Brooks (1981, 1985), in developing a method for studying host/parasite relations, developed an additive binary coding procedure which assumed that when several hosts were infected with one species of parasite, that group of hosts must be monophyletic. The analogous situation has been used in biogeography by a variety of authors (Wiley 1987; Wiley *et al.* 1991; Brooks and McLennan 1991). Zandee and Roos (1987) considered that widespread taxa should be considered monophyletic and no manipulation for areas was necessary, which in the spirit of Nelson and Platnick they described as assumption 0. Figure 9.15 shows the lizard example coded for assumption 0.

9.2.2.9 Comparing cladograms

Coding methods for turning trees into data matrices have been described by Brooks (1981), Wiley (1987), Zandee and Roos (1987), Brooks and McLennan (1991), and Wiley *et al.* (1991) (see Fig. 9.15). Algorithms for implementing assumptions 0, 1, and 2 have been written by Page (1989*a,b*). Page (1989*a*) points out that the implementation of assumptions 1 and 2 into the parsimony clique method for combining trees by Zandee and Roos (1987) is incomplete and so is not considered further here. The Wagner tree approaches of Kluge (1988*b*), Humphries *et al.* (1988), and Wiley (1987), and the parsimony-clique methods of Zandee and Roos (1987) require complex interpretations of homoplasy to account for the results. Attempts to apply parsimony to biogeography using programs designed for character analysis in systematics are similarly inappropriate. For example, equally parsimonious reconstructions of homoplasy means that different historical scenarios can be postulated for the same cladogram (Brooks 1985; Page 1989*a*).

Page (1989*a*) showed that the method of Nelson and Platnick (1981) for measuring the degree of parsimony of two cladograms is related to the algorithm

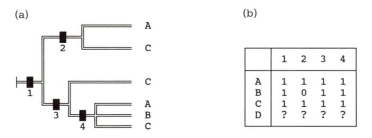

(a)   (b)

	1	2	3	4
A	1	1	1	1
B	1	0	1	1
C	1	1	1	1
D	?	?	?	?

Fig. 9.15 The lizard cladogram of Fig. 9.12a coded for further biogeographic analysis using Brooks Parsimony Analysis.

of Goodman *et al.* (1979) for reconciling gene and organism phylogenies and is also appropriate for comparing trees in biogeography. In Fig. 9.16a Nelson and Platnick interpret three possible solutions for resolving the widespread taxon in areas A and B. There are 15 possible resolutions for four areas, using assumption 1, three equally parsimonious and 12 less than parsimonious. Following Page, examples of each are given in Fig. 9.16b,c. For the three parsimonious solutions there are two items of error (Fig. 9.16b) and for the 12 less than parsimonious solutions there are 10 items of error (Fig. 9.16c). In an evolutionary scenario for Fig. 9.16c it is possible for the pattern to be explained by three extinctions – in C and D and in the ancestor of A and B (component 4).

9.2.2.10 Similarity of trees

There are a variety of means available for comparing the relative similarity of trees. Page (1988, 1989*a,b*) has described and implemented methods for calculating components, triplets, pruned trees, and pairwise distances in his program COMPONENT.

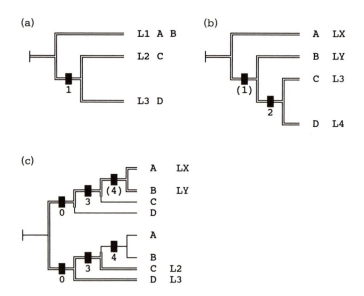

Fig. 9.16 (a) Cladogram for three species of lizard in areas A–D. (b) One of three equally parsimonious resolutions for four areas analysed under assumption 1. There are two items of error, component 1 and the extra taxon resulting from the splitting of L1 into Lx and Ly. (c) One of 12 less than parsimonious area cladograms. There are ten items of error, the missing components 0,3,3–4 and the missing taxa in areas A,B and C,D, component (4) and the extra taxon resulting from the splitting of L1 into Lx and Ly.

9.2.2.11 *General area cladograms; Nelson's consensus*

As distinct from the parsimony methods used for determining general area cladograms, as described in section 9.2.2.9 above, the degree as to which information in different cladograms agree can be found by using consensus trees. Nelson consensus (see also Chapter 5) is one appropriate device for determining the intersecting information in different cladograms. Consider the problem outlined in section 9.2.2.5 and Fig. 9.12. Under assumption 1 the occurrence of area C in all three clades can be reconciled by assuming that L1 has become extinct in area B and L2 has become extinct in areas A and B. The possibilities for further component analysis are given in Fig. 9.13e. Because the relations of areas A, B, and C are unresolved and there are four possible positions that area D can be added to the cladogram means that all 15 possible solutions for four taxa are recovered. The same thing applies to the frog and bird cladograms, which means that all possible intersections are possible in the Nelson consensus tree.

Under assumption 2 however, the one resolved three-area statement, (A(B C)), provides five possibilities for further component analysis (Fig. 9.17a). Addition of area D can be applied to five possible positions as summarized in Fig. 9.17b. For the frog and bird cladograms there are two resolved three-area statements which each provides ten possibilities for further analysis by adding the 'missing' area A (Fig. 9.17c). All 15 possible cladograms are recovered when looking at the total 25 cladograms determined by adding the possible positions that the 'missing' areas might occupy but the intersections provide an interesting result (Fig. 9.17d). Comparing across the three sets of possible cladograms shows six unique patterns, four for the bird cladogram and two for the frog cladogram, eight intersections for two of the cladograms, four between frog and bird cladograms, three between the lizard and frog cladograms, one between the lizard and bird cladograms, and finally only one intersection across all three cladograms, the pattern (A(B(C D))) (see also Fig. 9.12d). This pattern is recovered in the Nelson consensus tree because it is the maximum clique when adding all 25 possible trees together (Fig. 9.17d).

Comparing the results with the original area cladograms (Fig. 9.12a–c) indicates which areas transmit the historical signal and which are noise. For example, in the lizard cladogram areas A for L1, B for L3, and C for L2 transmit the signal and areas C for L1, and A and C in L3 are the noise.

9.2.2.12 *Platnick's example (1981); Page's method (1988, 1989a,b)*

Returning to the two poeciliid fish genera *Heterandria* and *Xiphophorus* each has 11 identifiable disjunct areas. The area cladograms are shown in Fig. 9.18, calculated for assumptions 0, 1 and 2 with the program COMPONENT (Page 1988, 1989a,b). Areas 4 and 5 are occupied by one species in each genus and are thus treated as a single area. Area 11 was treated by Rosen (1978, 1979) as a putatively hybrid area between areas 4, 5, and 2 and, following Platnick (1981) and Page (1988, 1989a,b), it is left out of the analysis.

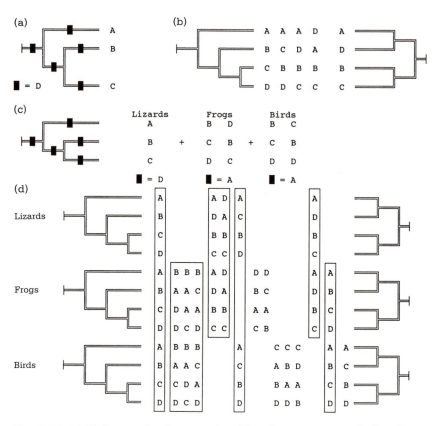

Fig. 9.17 (a) Cladogram for three species of lizard in areas A–D; the bar shows possible positions of 'missing' area D. (b) One of three equally parsimonious resolutions for four areas analysed under assumption 1 showing positions for the 'missing' area D; the five possible cladograms obtained by adding area D to Fig. 9.17a. (c) Five possible three-area cladograms obtained by analysis with assumption 2. (d) Intersecting cladograms obtained by assumption 2 and addition of 'missing' areas to the cladograms in Fig. 9.17b. The lizards, frogs and birds display a variety of unique or partial intersections but only one, (A(B(C D))), that is found in all three.

A comparison of the two cladograms (Fig. 9.18) shows that *Xiphophorus* is less informative than *Heterandria* because it has two widespread species in areas 4, 5, 6, 9, and 10 and is absent from area 7. In *Heterandria* areas 4, 5, 6, 9, and 10 are all occupied by recognizable endemic taxa.

As stated earlier, under assumption 1 whatever is true of a widespread taxon in one part of its range (for example *Xiphophorus alvarezi* in area 4, 5) must also be true in the other part of its range (i.e. area 6). However, under assumption 2 whatever is true of a widespread taxon in one part of its range need not also be true of the taxon elsewhere. In other words, the widespread distributions

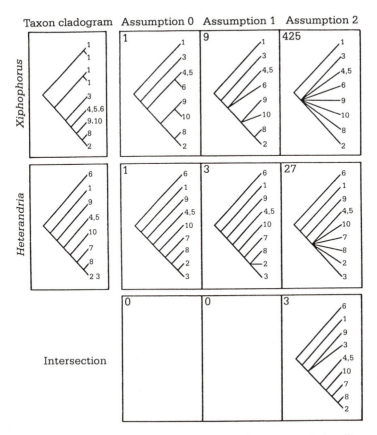

Fig. 9.18 Reanalysis of Rosen's (1978, 1979) data. For *Xiphophorus* and *Heterandria* the taxon cladogram and Nelson and Majority rule consensus trees analysed under assumptions 0, 1 and 2 are shown. The number in each box is the number of equally parsimonious area cladograms. The intersection is the set of area cladograms that both *Xiphophorus* and *Heterandria* support (redrawn from Page 1988).

are equivalent to saying that we are ignorant of the reasons for lack of resolution in the cladogram. In distributional terms, that is equivalent to saying that we do not know whether the patterns are due to dispersal or a failure to speciate in response to a vicariance event. Rosen's original application of the Platnick and Nelson (1978) biogeographic method in 1978 and 1979 compared the two clado-grams to one another and identified only those parts which were congruent. By pruning away the incongruent and unique areas a cladogram for only six areas could be produced. Figure 9.18 shows two area cladograms for *Heterandria* and *Xiphophorus*. Platnick (1981) considered the removal of unique and incon-gruent areas equivalent to analysing under assumption 1. If assumption 1 is adopted, then the *Xiphophorus* populations of area 9 must be most closely related to the population in area 10, and the information for area 9 is incongruent with the information from *Heterandria*. Similarly, the information on area 6 is also incongruent to both cladograms. Hence the nine area cladograms for *Xiphophorus* and the three cladograms for *Heterandria* have no intersections (Fig. 9.18).

A different result can be obtained by applying assumption 2. By taking the information on areas 6 and 9 from *Heterandria* as correct then the incongruent information in the same areas for *Xiphophorus* is due either to dispersal or a failure to speciate in response to a vicariance event. Rosen's original reasons for applying a pruning algorithm was that when groups dispersed or failed to respond to vicariance events it reduced informativeness. Platnick (1981) noted however, that if widespread taxa are uninformative they cannot at the same time be incongruent. Absence data can never be incongruent with information at hand so unique areas should never be deleted on these grounds. Taken on their own, widespread taxa under assumption 2 give uninformative components but when considered with other cladograms involving widespread taxa informative results are possible. Under assumption 2 the *Xiphophorus* cladogram (Fig. 9.18) allows the populations in area 9 or 10 (but not both) and area 4, 5, or 6 (but not all) to occur in any of 12 positions, on nearly all branches of the cladogram. Of the 425 possible trees for *Xiphophorus* and the 27 for *Heterandria* the analysis yields three possible intersections (using Nelson or Majority rule consensus). These can be summarized in a consensus tree by a trichotomy as shown by an asterisk in Fig. 9.18.

We have a cladistic structure that accounts for all ten areas of endemism (excluding the hybrid area 11) that can be recognized from the two fish genera *Heterandria* and *Xiphophorus*. This gives an almost identical result for the ten areas to that obtained by Wiley's (1980, 1981) method, but obtained by using cladogram logic rather than evolutionary assumptions. If such a pattern is due to changes in Earth history the question that we could ask now is what might have been the historical factors in Mesoamerica to cause this pattern and how might these be compared with the given biological distribution? So that biotic and historical patterns can be compared we would ideally require that geological information be assembled into cladograms in the same way as biological cladograms. Until such time as geological data can be ordered for a more informative comparison one can say little except that the observed patterns in Mesoamerica have been formed over a period of at least the last 80 million years (Rosen 1978).

9.3 CONCLUSIONS

At this time, methods of cladistic biogeography are very useful for analysing and comparing biotic patterns at the highest resolution so as to compare them to independent sources of data such as geological patterns. Furthermore, cladistics is a general method of determining class and subclass relations, whatever the source of data, without recourse to evolutionary narrative (see Nelson 1982). A cladistic view of world history combined with the cladistic method in systematics makes it possible to express area interrelationships as hierarchical relations from biotic information. The development of methods, such as Nelson and Platnick's assumption 2, allows the possibility of generating general biogeographic hypotheses from congruent cladograms even from seemingly ambiguous patterns. Assumption 2 is a general empirical procedure without any dispersal, vicariant or extinction events assumed in the analysis but which at the same time never denies that they occur (Nelson 1982).

10.
Formal classification

Peter L. Forey

10.1 INTRODUCTION

Classification is the 'activity of grouping entities or phenomena and naming the resultant groupings' (Wiley 1979, p. 309). As such, the naming of groups and formal listing of groups are intimately connected with the method of grouping. But the subject matter here is concerned with the translation of the cladogram (or phylogenetic tree) into words (or numbers) and, for want of an explicit term, this activity is called formal classification.

10.2 HIERARCHIES AND NATURAL GROUPS

Formal classification is hierarchical, and most systematists have worked within the framework of the Linnaean hierarchy with the familiar rank names (Genus, Kingdom, etc.). There also seems to be some consensus of opinion that formal classifications should be natural, i.e. made up of natural groups. However, the arguments centre on what is meant by natural. Cladists are clear on this point. Natural groups are monophyletic groups and therefore formal classification should contain monophyletic groups in so far as this is possible. Fossil groups present some problems and any deviation from monophyly should be noted by convention.

For evolutionary systematists the issue is different and not so clear. Evolutionary taxonomists recognize two kinds of relationship for classification − genetic and genealogical (see also Chapter 1). In these instances the diagram (tree) has to be read alongside the written classification. Bock (1974) sums up the view by suggesting that the diagram and formal classification should be mirror images of one another. By this he means that whichever of the two relationships is not shown at a particular point on the tree should be expressed in the corresponding part of the formal classification. A familiar example would be a tree showing crocodiles more closely related, genealogically, to birds and in which the formal classification includes crocodiles with snakes, lizards, and turtles as the Reptilia. This would be done to reflect the view that reptiles (including crocodiles) occupy an adaptive zone quite different to that of birds. Thus for many evolutionary taxonomists, there can be an asymmetry between the tree and the formal classification, and indeed this is considered desirable. It also means that the formal classification would contain groups (Aves, Mammalia) which would have characters and non-groups (Reptilia, Pisces) which do not. Cladists find it difficult to understand what is gained by retaining these non-groups; such activity seems merely to be an appeal to tradition. However, many evolutionary taxonomists share Darwin's wish that classifications will become genealogical and that the ideal classification would consequently match the tree.

10.3 NAMES OF GROUPS

Cladists accept that, if names are applied to groups, those names should convey a precise message to the reader. After all, formal classification is an international means of communication. This communication may take place between a systematist, who presumably understands the classification, and say a physiologist who may not understand how the classification was constructed but wishes to interpret his or her data in a comparative way. So cladists believe that formal classifications should reflect the cladogram. The problem therefore becomes one of how to translate the cladogram or phylogenetic tree into words without ambiguity.

Several systems of words and numbers have been devised to deal specifically with cladograms and phylogenetic trees. Only a few are described here but fuller accounts can be found in Wiley (1979) and Bonde (1977).

10.3.1 Equivalent ranks

One of the principal suggestions offered by Hennig was that if formal classification is to reflect hypotheses of relationship accurately then sister taxa ought to be of equivalent rank (Fig. 10.1). This recommendation has been followed by most cladists but there are certain problems where the cladogram is very asymmetrical. For instance, in Fig. 10.2 there are six taxa which, following Hennig's recommendation, would require the use of five ranks (excluding species). This situation arises particularly with the inclusion of fossil stem species. In the theoretical example species A is included within three 'empty ranks', Genus, Family, and Order: i.e. those ranks would be superfluous because they convey no more information than is already contained in the ranks Species and Class. 'Empty ranks' arise in traditional classifications because the particular taxon is thought to be different, morphologically, that high rank is required to convey information. In other words, rank is tied to importance of characters.

It is clear that a large number of ranks might be needed in groups with many species. This problem has prompted several solutions. McKenna (1975) attempted a formal classification of fossil and Recent mammals in which he used

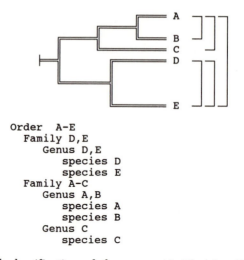

```
Order  A-E
  Family D,E
    Genus D,E
        species D
        species E
  Family A-C
    Genus A,B
        species A
        species B
    Genus C
        species C
```

Fig. 10.1 Formal classification of the group (A–E) using Hennig's (1966) convention of naming sister groups at equivalent rank.

many levels of rank names with conventional prefixes (for example Superlegion, Infralegion, etc.) together with many new names for the various groupings. Farris (1976) on the other hand invented a series of prefix combinations which could be attached to familiar ranks (for example mega/hyper/super-cohort) to provide a very large number of ranks. Other authors (for example Hennig 1969; Griffiths 1974; Løvtrup 1977) have suggested that a numerical rank system might be used (Figs 10.3 and 10.4). As long as only monophyletic groups are used such numerical systems are internally consistent but they suffer the disadvantages of being very unfriendly and unfamiliar.

10.3.2 Indentation and subordination

Another system is one of indenting the names of taxa to correspond with the system of subordination suggested by the cladogram. In such a system the taxa may be retained with the traditional ranks, since it is not the rank which is important, only its position on the page. Sister taxa are indented by the same amount. Farris (1976) has discussed the indenting system. Wiley (1979) points out that indented systems suffer from practical difficulties where classifications are long and extend over several pages because it may be difficult to recognize co-ordinate taxa except by measuring the distance between the word and the page margin.

To avoid the problem of running out of ranks we could present incomplete classifications. For instance, we could classify down to a certain level in a particularly species-rich insect group and then simply list the genera, leaving the

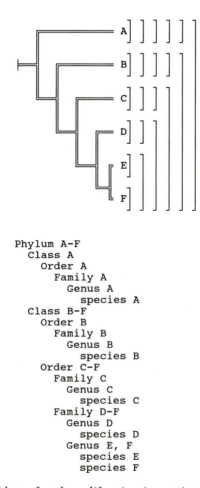

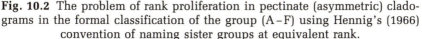

```
Phylum A-F
   Class A
      Order A
         Family A
            Genus A
               species A
   Class B-F
      Order B
         Family B
            Genus B
               species B
      Order C-F
         Family C
            Genus C
               species C
         Family D-F
            Genus D
               species D
            Genus E, F
               species E
               species F
```

Fig. 10.2 The problem of rank proliferation in pectinate (asymmetric) clado-grams in the formal classification of the group (A–F) using Hennig's (1966) convention of naming sister groups at equivalent rank.

diagram to convey the theory of relationship. This is, after all, done in most classifications.

10.3.3 Sequencing

Nelson (1972, 1973) has suggested another convention – sequencing. This consists of listing taxa in a sequence, such that each taxon is the sister group of that succeeding it (Fig. 10.5). Nelson was prompted to recommend this practice because of certain problems created by inclusion of fossils in a formal classification designed for Recent organisms.

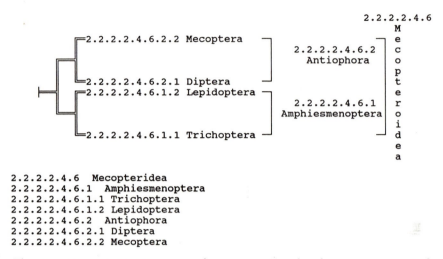

2.2.2.2.4.6   Mecopteridea
2.2.2.2.4.6.1   Amphiesmenoptera
2.2.2.2.4.6.1.1   Trichoptera
2.2.2.2.4.6.1.2   Lepidoptera
2.2.2.2.4.6.2   Antiophora
2.2.2.2.4.6.2.1   Diptera
2.2.2.2.4.6.2.2   Mecoptera

Fig. 10.3 Hennig's (1966) system of representation of ranks using a numerical classification numbering from the terminals to the root of the classification.

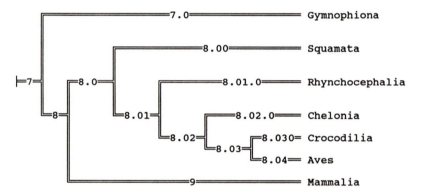

Fig. 10.4 Løvtrup's (1977) system of representation of ranks using a numerical classification numbering from the root to the terminals of the classification.

10.4 FOSSILS

If only phylogenetic trees are considered then some fossil species may be thought of as ancestors, albeit that such ideas have to be treated in special ways. The problem caused by this idea is that the ancestor is co-extensive with its presumed descendants. This means that, for formal classification, the ancestor = group (for example *Archaeopteryx* = class Aves). Hennig (1966) assigned rank

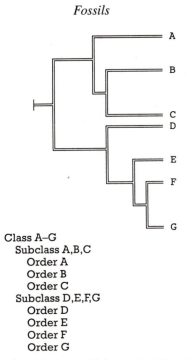

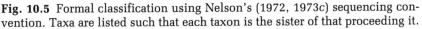

Class A–G
  Subclass A,B,C
    Order A
    Order B
    Order C
  Subclass D,E,F,G
    Order D
    Order E
    Order F
    Order G

Fig. 10.5 Formal classification using Nelson's (1972, 1973c) sequencing convention. Taxa are listed such that each taxon is the sister of that proceeding it.

according to the time of origin of the group. In instances in which ancestors are recognized this means that both ancestor and group occurred at the same time. Hence, they must receive the same rank. The solution to this problem is to regard all taxa, whether they be ancestors or not, as terminal taxa (Patterson and Rosen 1977) and not recognize ancestors at all. Wiley (1979) took a slightly different view because he is concerned with phylogenetic trees. He accepts that it might be possible to recognize ancestral species and that it would be beneficial to include this information in the formal classification. Since the ancestral species is co-extensive with the group to which it belongs, then it can be listed in parentheses following the group name. This convention is applicable only to phylogenetic trees, cladograms do not contain ancestors.

10.4.1 Fossils and Recent groups

A second problem concerns the fact that fossils and fossil groups are, more often than not, plesiomorphic sister groups of Recent groups. As stem lineages become better known and interpreted using cladistic analysis it is often possible to resolve them as a series of more inclusive sister groupings (see Patterson 1977; Patterson and Rosen 1977). This produces asymmetrical cladograms or phylogenetic trees where the terminal taxa are often single species. The problems of rank and the profusion of required names for described groups was described earlier. The problem would be exacerbated should a new fossil be found which could be interpolated along the stem lineage series. For some authors (for example Crowson 1970) attempts to fit fossils into classifications of Recent organisms create so many problems that separate classifications should be used. Crowson suggested that separate classifications should be erected for arbitrarily selected units of time (for example the Cretaceous). Patterson and Rosen (1977) give a practical example of this suggestion and point out the difficulties which follow. Here it is assumed that most workers wish to combine Recent and fossil classifications.

10.4.2 Sequences and plesions

The sequencing technique avoids such problems where fossil taxa are merely listed in sequence, reflecting the geometry of the cladogram or phylogenetic tree. Nelson suggested that these fossil taxa (representing stem-group species) all be given the same rank as Recent sister-groups and be denoted with a special symbol such as a dagger. The major problem with this convention is that fossil taxa (which in some cases are only single specimens) may have very high rank, dependent on the accepted rank of the 'crown group'. So Patterson and Rosen (1977) modified Nelson's suggestion by introducing a new rank name – plesion – for each of these fossil taxa (Fig. 10.6). The concept is explained in detail by Patterson and Rosen (1977) where they give examples and where they also denote the species or group with a dagger symbol. Strictly speaking this is unnecessary because the plesion is, by definition, extinct.

Patterson and Rosen (1977), Nelson (1972), and Wiley (1979) all discuss the inclusion of fossil paraphyletic groups, fossils in which few characters are known and the presentation of trichotomies, etc. These apparently difficult situations can be considered together because, in each case, some indication of uncertainty has to be implied. Fossil paraphyletic groups can be indicated by including the taxon in quotes. Patterson and Rosen (1977) also use the same convention for those groups from which the type genus has been removed and they give an example where the quotes mean both. It is clear that the phylogenetic position of such 'groups' cannot be accurately specified and has to be used with another convention – *incertae sedis*. Nelson (1972) and Patterson and Rosen (1977) use *incertae sedis* for those fossil groups whose lower position cannot be accurately assessed. This may happen because of lack of information (incomplete fossils) and signifies that a particular fossil group is interchangeable with another in the cladogram. The taxon in question is placed *incertae sedis* in the most inclusive group to which it can be assigned. For example, Patterson and Rosen (1977) list the fossil teleostean genus *Anaethalion* as Elopocephala *incertae sedis*. This means that *Anaethalion* shares a synapomorphy of the supercohort Elopocephala but no other synapomorphies of any elopocephalan sub-group.

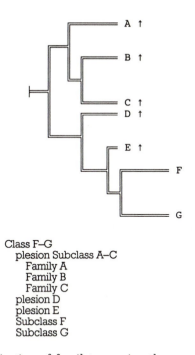

Class F–G
    plesion Subclass A–C
        Family A
        Family B
        Family C
    plesion D
    plesion E
    Subclass F
    Subclass G

Fig. 10.6 The classification of fossil taxa using the conventions of plesions (Patterson and Rosen 1977) and sequencing.

The restriction of the *incertae sedis* category to fossil taxa may seem unusual (Wiley rejects the action as unnecessary) but it was recommended by Nelson and by Patterson and Rosen to signify that the uncertainty exists because of a lack of information. Uncertainty about the phylogenetic position of a Recent taxon arises from deficient theories of relationship (lack of information is not a problem of Recent taxa − just of the investigator). Uncertain relationships among Recent organisms (customarily designated by the *incertae sedis* convention) can be indicated instead by listing more than two equally ranked taxa within the same group. On the cladogram this would be represented as a polytomy. This is perfectly satisfactory for a formal classification of a cladogram which is concerned only with the distribution of synapomorphies and in which polytomies are expressions of ignorance. But for phylogenetic trees the listing of more than two Recent taxa could mean additionally, a precise statement of polytomous speciation or an indication of hybridization or reticulate evolution. Wiley (1979) adopts a convention for hybridization (listing the presumed parents against the name) and for the 'expression of ignorance' he chooses to place the phrase *sedis mutabilis* (meaning changeable position) following each of the taxa in the polytomy. Wiley would, however, use the *sedis mutabilis* suffix for fossil groups (including plesions) as well, but this seems unnecessary. Examples of *incertae sedis*, plesion and *sedis mutabilis* are given in Fig. 10.7.

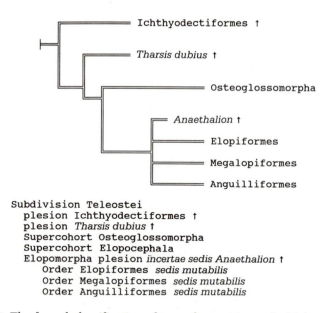

Subdivision Teleostei
  plesion Ichthyodectiformes †
  plesion *Tharsis dubius* †
  Supercohort Osteoglossomorpha
  Supercohort Elopocephala
  Elopomorpha plesion *incertae sedis Anaethalion* †
      Order Elopiformes *sedis mutabilis*
      Order Megalopiformes *sedis mutabilis*
      Order Anguilliformes *sedis mutabilis*

Fig. 10.7 The formal classification of taxa, the positions of which are poorly understood, using the concepts of uncertain position (*incertae sedis*), change-able position (*sedis mutabilis*), and fossils (plesions).

10.5 SUMMARY

To summarize, it is possible to translate accurately a cladogram or phylogenetic tree of Recent organisms into a formal classification without loss of information by using conventions of subordination and/or sequencing. For cladograms no further convention is required, since a cladogram is a statement about the distribution of synapomorphies which enable natural groups to be recognized. Paraphyletic assemblages, polytomies, are denoted by listing more than two equally ranked taxa within a more inclusive grouping. Cladograms which include fossil species can use the additional conventions of plesion and *incertae sedis*. For phylogenetic trees of fossil and Recent organisms, which include evolutionary concepts such as ancestors, polychotomous speciation events, hybridization, taxa of symbiotic origin then several more conventions have to be added but Wiley (1979) has provided a suite of annotations to enable this to be undertaken.

   The techniques described above are designed primarily to denote relative rank, but Hennig did discuss the possibility of according absolute rank. He did not, however, consider it necessary for his phylogenetic system (1966; p. 191), but suggested that if one considered a genus of flies, then this ought to be comparable in some way with a genus of birds. This appears to be a traditional viewpoint. It is also necessary to remember that Hennig was concerned more about phylogenetic trees, rather than with cladograms, and he thought that the assignment of absolute rank, reflecting age of origin, would complete the information in the tree.

In older, traditional and typological classifications, absolute ranks was associated with characters. For instance, in brachiopods, Muir-Wood (1955) mentions generic/specific characters and superfamily/order characters. Whether a genus is placed in an existing family or another depends on subjective evaluation of divergence. Hennig suggested that a non-subjective method of assigning rank might be to relate rank to the age of the group. Hennig divided the Phanerozoic into time bands (1966; Fig. 58). Thus, any group which originated between the Upper Carboniferous and Upper Permian (Hennig's time band III) would be accorded ordinal rank. The absolute age of origin of a group is determined by reference to the fossil record (hence the importance Hennig placed on fossils), biogeographical information associated with theories of Earth history, and a parasitological method. Hennig made a distinction between the age of origin and the age of differentiation (crown group) of a group (see Chapter 8; Fig. 8.2). He recognized that there may be a significant time difference between the two, and that this period may have been occupied by a series of fossil species constituting a stem-group series. The 'empty ranks' mentioned above may be taken up by these species, with the necessary ranks (highest and lowest) denoting the time of origin and time of differentiation, respectively (see Chapter 8).

Farris (1974) suggested a different system of recognizing absolute rank, in which he ascribed rank of a Recent group based on the age of differentiation and rank of the fossil group based on longevity of the group. This is a device to avoid the use of empty ranks.

The attempts to assign absolute ranks to groups based on either the time of origin or time of differentiation has not been followed up in recent years. This is probably because cladograms (with no absolute time dimension) have become more important than phylogenetic trees and other conventions such as plesions and sequencing have been adopted. Further consideration of both Hennig's and Farris' ideas may be worthwhile for those interested in phylogenetic trees.

References

Aboitiz, F. (1988). Homology, a comparative or a historical concept? *Acta Biotheoretica*, **37**, 27–9.

Adams, E. N. (1972). Consensus techniques and the comparison of taxonomic trees. *Systematic Zoology*, **21**, 390–97.

Alberch, P. (1985). Problems with the interpretation of developmental sequences. *Systematic Zoology*, **34**, 46–58.

Alberch, P., Gould, S. J., Oster, G. F., and Wake, D. B. (1979). Size and shape in ontogeny and phylogeny. *Paleobiology*, **5**, 296–317.

Anderberg, A. and Tehler, A. (1990). Consensus trees, a necessity in taxonomic practice. *Cladistics*, **6**, 399–402.

Archie, J. W. (1989). Homoplasy Excess Ratios: new indices for measuring levels of homoplasy in phylogenetic systematics and a critique of the Consistency Index. *Systematic Zoology*, **38**, 253–69.

Arnold, E. N. (1981). Estimating phylogenies at low levels. *Zeitschrift für zoologische Systematik und Evolutionsforschung*, **19**, 1–35.

Ax, P. (1987). *The phylogenetic system*. John Wiley, Chichester.

Bailey, I. W. (1949). Origin of the angiosperms: need for a broadened outlook. *Journal of the Arnold Arboretum Harvard University*, **30**, 64–70.

Ball, I. R. (1976). Nature and formulation of biogeographic hypotheses. *Systematic Zoology*, **24**, 407–30.

Beatty, J. (1982). Classes and cladists. *Systematic Zoology*, **31**, 25–34.

Beyer, W. A., Stein, M. L., Smith, T. F., and Ulman, S. M. (1974). A molecular sequence metric and evolutionary trees. *Mathematical Biosciences*, **19**, 9–25.

Bishop, M. J. (1982). Criteria for the determination of the direction of character state changes. *Zoological Journal of the Linnean Society*, **74**, 197–206.

Bishop, M. J. and Friday, A. E. (1985). Evolutionary trees from nucleic acid and protein sequences. *Proceedings of the Royal Society London*, B, **226**, 271–302.

Blackmore, S. (1986). Cellular ontogeny. *Cladistics*, **2**, 358–61.

Bock, W. (1974). Philosophical foundations of evolutionary classification. *Systematic Zoology*, **22**, 375–92.

Bock, W. J. (1989). The homology concept: its philosophical and practical methodology. *Zoologische Beiträge, NF*, **32**, 327–53.

Bonde, N. (1977). Cladistic classification as applied to vertebrates. In *Major patterns in vertebrate evolution* (ed. M. K. Hecht, P. C. Goody, and B. M. Hecht), pp. 741–804. Plenum Press, New York.

Bonheim, H. (1990). *Literary systematics*. D. S. Brewer, Cambridge.

Brady, R. (1982). Theoretical issues and 'pattern cladists'. *Systematic Zoology*, **31**, 286–91.

Bremer, K. (1990). Combinable component consensus. *Cladistics*, **6**, 369–72.

Bretsky, S. S. (1975). Allopatry and ancestors: a response to Cracraft. *Systematic Zoology*, **24**, 113–19.

Britten, R. J. (1986). Rates of sequence evolution differ between taxonomic groups. *Science*, **231**, 1393–8.

Brooks, D. R. (1981). Hennig's parasitological method: a proposed solution. *Systematic Zoology*, **30**, 229–49.

Brooks, D. R. (1985). Historical ecology: A new approach to studying the evolution of ecological associations. *Annals of the Missouri Botanical Garden*, **72**, 660–80.

Brooks, D. R. and McLennan, D. A. (1991). *Phylogeny, ecology, and behavior: a research program in comparative biology*. University of Chicago Press, Chicago.

Brooks, D. R. and Wiley, E. O. (1985). Theories and methods in different approaches to phylogenetic systematics. *Cladistics*, **1**, 1–11.

Brown, W. M., Praeger, E. M., Wang, A., and Wilson, A. C. (1982). Mitochondrial DNA sequences of primates: tempo and mode of evolution. *Journal of Molecular Evolution*, **18**, 225–39.

Brundin, L. (1966). Transantarctic relationships and their significance as evidenced by midges. *Kungliga Svenska Vetenskapsakademiens Handlinger*, **4**, (11), 1–472.

Brundin, L. (1972a). Evolution, causal biology, and classification. *Zoologica Scripta*, **1**, 107–20.

Brundin, L. (1972b). Phylogenetics and biogeography. *Systematic Zoology*, **21**, 69–79.

Brundin, L. (1981). Croizat's biogeography versus phylogenetic biogeography. In *Vicariance biogeography: a critique* (ed. G. Nelson and D. E. Rosen), pp. 94–158. Columbia University Press, New York.

Camin, J. H. and Sokal, R. R. (1965). A method for deducing branching sequences in phylogeny. *Evolution*, **19**, 311–26.

Carpenter, J. M. (1988). Choosing among multiple equally parsimonious cladograms. *Cladistics*, **4**, 291–6.

Carpenter, J. M. (1990). Of genetic distances and social wasps. *Systematic Zoology*, **39**, 391–7.

Cavender, J. A. (1989). Mechanized derivation of linear invariants. *Molecular Biology and Evolution*, **6**, 301–16.

Chappill, J. A. (1989). Quantitative characters in phylogenetic analysis. *Cladistics*, **5**, 217–34.

Clark, C. and Curran, D. J. (1986). Outgroup analysis, homoplasy, and global parsimony: a response to Maddison, Donoghue and Maddison. *Systematic Zoology*, **35**, 422–6.

Clarke, C. A. (1980). *Papilio nandina*, a probable hybrid between *Papilio dardanus* and *Papilio phorcas*. *Systematic Entomology*, **5**, 49–57.

Cloutier, R. (1991). Interrelationships of Palaeozoic actinistians: patterns and trends. In *Early vertebrates and related problems of evolutionary biology* (ed. M.-M. Chang, Y.-H. Liu, and G. Zhang), pp. 379–428. Science Press, Beijing.

Cracraft, J. (1979). Phylogenetic analysis, evolutionary models and paleontology. In *Phylogenetic analysis and palaentology* (ed. J. Cracraft and N. Eldredge), pp. 7–39. Columbia University Press, New York.

Cranston, P. S. and Humphries, C. J. (1988). Cladistics and computers: a chironomid conundrum? *Cladistics*, **4**, 72–92.

Crisci, J. and Stuessy, T. (1980). Determining primitive character states for phylogenetic reconstruction. *Systematic Botany*, **6**, 112–35.

Croizat, L. (1952). *Manual of phytogeography*. W. Junk, The Hague.

Croizat, L. (1958). *Panbiogeography*. Published by the author, Caracas.

Croizat, L. (1964). *Space, time, form: the biological synthesis*. Published by the author, Caracas.

Croizat, L., Nelson, G. J., and Rosen, D. E. (1974). Centers of origin and related concepts. *Systematic Zoology*, **23**, 265–87.

Cronquist, A. (1988). A botanical critique of cladism. *The Botanical Review*, **53**, 1–52.

Crowson, R. A. (1970). *Classification and biology*. Atherton Press, New York.

Darlu, P. and Tassy, P. (1987). Roots (a comment on the evolution of human mitochondrial DNA and the origins of modern humans). *Human Evolution*, **2**, 407–12.

Davis, P. H. and Heywood, V. H. (1963). *Principles of angiosperm taxonomy*. Oliver & Boyd, Edinburgh.

De Soete, G. (1983). On the construction of optimal phylogenetic trees. *Zeitschrift für Naturforschung*, **38c**, 156–8.

DeBry, R. W. and Slade, N. A. (1985). Cladistic analysis of restriction endonuclease cleavage maps within a maximum-likelihood framework. *Systematic Zoology*, **34**, 21–34.

De Queiroz, K. (1985). The ontogenetic method for determining character polarity and its relevance to phylogenetic systematics. *Systematic Zoology*, **34**, 280–99.

Devereux, R., Loehlich, A. R., and Fox, G. E. (1990). Higher plant origins and the phylogeny of green algae. *Journal of Molecular Evolution*, **31**, 18–24.

Dobson, A. J. (1974). Unrooted trees in numerical taxonomy. *Journal of Applied Probability*, **11**, 32–42.

Dollo, L. (1893). Les lois de l'évolution. *Bulletin de la Société Belge de Géologie, de Paléontologie et d'Hydrologie*, **7**, 164–6.

Donoghue, M. J. and Cantino, P. D. (1984). The logic and limitations of the outgroup substitution approach to cladistic analysis. *Systematic Botany*, **9**, 192–202.

Donoghue, M. J., Doyle, J. A., Gauthier, J., Kluge, A. G., and Rowe, T. (1989). The importance of fossils in phylogeny reconstruction. *Annual Review of Ecology and Systematics*, **20**, 431–60.

Doyle, J. A. and Donoghue, M. J. (1986). Seed plant phylogeny and the origin of the angiosperms: an experimental approach. *The Botanical Review*, **52**, 321–431.

Doyle, J. A. and Donoghue, M. J. (1987). The importance of fossils in elucidating seed plant phylogeny and macroevolution. *Reviews in Palaeobotany and Palynology*, **50**, 63–95.

Edmunds, J. (1965). Paths, trees and flowers. *Canadian Journal of Mathematics*, **17**, 449–67.

Eldredge, N. (1979). Alternative approaches to evolutionary theory. *Bulletin Carnegie Museum of Natural History*, **13**, 7–19.

Eldredge, N. and Cracraft, J. (1980). *Phylogenetic patterns and the evolutionary process*. Columbia University Press, New York.

Estabrook, G. F., Johnson, C. S. Jr., and McMorris, F. R. (1976a). An algebraic analysis of cladistic characters. *Mathematical Biosciences*, **29**, 181–7.

Estabrook, G. F., Johnson, C. S. Jr., and McMorris, F. R. (1976b). A mathematical foundation for the analysis of cladistic character compatibility. *Discrete Mathematics*, **16**, 141–7.

Faith, D. P. (1985). Distance methods and the approximation of most-parsimonious trees. *Systematic Zoology*, **34**, 312–25.

Faith, D. P. (1991). Cladistic permutation for monophyly and nonmonophyly. *Systematic Zoology*, **40**, 366–75.

Faith, D. P. and Cranston, P. S. (1990). Could a cladogram this short have arisen by chance alone?: on permutation tests for cladistic structure. *Cladistics*, **7**, 1–28.

Farris, J. S. (1969). A successive approximations approach to character weighting. *Systematic Zoology*, **18**, 374–85.

Farris, J. S. (1970). Methods for computing Wagner trees. *Systematic Zoology*, **19**, 83–92.

Farris, J. S. (1972). Estimating phylogenetic trees from distance matrices. *American Naturalist*, **106**, 645–68.

Farris, J. S. (1974). Formal definitions of paraphyly and polyphyly. *Systematic Zoology*, **23**, 548–54.

Farris, J. S. (1976). Phylogenetic classification of fossils with recent species. *Systematic Zoology*, **25**, 271–82.

Farris, J. S. (1977*a*). On the phenetic approach to vertebrate classification. In *Major patterns in vertebrate evolution* (ed. M. K. Hecht, P. C. Goody and B. M. Hecht), pp. 823–50. Plenum, New York.

Farris, J. S. (1977*b*). Phylogenetic analysis under Dollo's Law. *Systematic Zoology*, **26**, 77–88.

Farris, J. S. (1978). Inferring phylogenetic trees from chromosome inversion data. *Systematic Zoology*, **27**, 275–84.

Farris, J. S. (1979*a*). On the naturalness of phylogenetic classification. *Systematic Zoology*, **28**, 200–13.

Farris, J. S. (1979*b*). The information content of the phylogenetic system. *Systematic Zoology*, **28**, 483–519.

Farris, J. S. (1980). The efficient diagnoses of the phylogenetic system. *Systematic Zoology*, **29**, 386–401.

Farris, J. S. (1981). Distance data in phylogenetic analysis. *Advances in Cladistics*, **1**, 3–23.

Farris, J. S. (1982*a*). Outgroups and parsimony. *Systematic Zoology*, **31**, 328–34.

Farris, J. S. (1982*b*). Simplicity and informativeness in systematics and phylogeny. *Systematic Zoology*, **31**, 413–44.

Farris, J. S. (1983). The logical basis of phylogenetic analysis. *Advances in Cladistics*, **2**, 7–36.

Farris, J. S. (1985). Distance data revisisted. *Cladistics*, **1**, 67–85.

Farris, J. S. (1986*a*). On the boundaries of phylogenetic systematics. *Cladistics*, **2**, 14–27.

Farris, J. S. (1986*b*). Distances and statistics. *Cladistics*, **2**, 144–57.

Farris, J. S. (1988). *Hennig86 version 1.5 manual*; software and MSDOS program.

Farris, J. S. (1989). The retention index and the rescaled consistency index. *Cladistics*, **5**, 417–19.

Farris, J. S. (1991). Excess homoplasy ratios. *Cladistics*, **7**, 81–91.

Farris, J. S., Kluge, A. G., and Eckhardt, M. J. (1970). A numerical approach to phylogenetic systematics. *Systematic Zoology*, **19**, 172–89.

Felsenstein, J. (1978*a*). The number of evolutionary trees. *Systematic Zoology*, **27**, 27–33.

Felsenstein, J. (1978*b*). Cases in which parsimony and compatibility methods will be positively misleading. *Systematic Zoology*, **27**, 401–10.

Felsenstein, J. (1979). Alternative methods of phylogenetic inference and their interrelationship. *Systematic Zoology*, **28**, 49–62.

Felsenstein, J. (1981*a*). A likelihood approach to character weighting and what it tells us about parsimony and compatibility. *Biological Journal of the Linnean Society*, **16**, 183–96.

Felsenstein, J. (1981*b*). Evolutionary trees from DNA sequences: a maximum likelihood approach. *Journal of Molecular Evolution*, **17**, 368–76.

Felsenstein, J. (1982). Numerical methods for inferring evolutionary trees. *Quarterly Review of Biology*, **57**, 379–404.

Felsenstein, J. (1983). Statistical inference of phylogenies. *Journal of the Royal Statistical Society*, A, **146**, 246–72.

Felsenstein, J. (1984). Distance methods for inferring phylogenies: a justification. *Evolution*, **38**, 16–24.

Felsenstein, J. (1985). Confidence limits on phylogenies: an approach using the bootstrap. *Evolution*, **39**, 783–91.

Felsenstein, J. (1986). Distance methods: a reply to Farris. *Cladistics*, **2**, 130–44.

Felsenstein, J. (1988*a*). Phylogenies and quantitative characters. *Annual Review of Ecology and Systematics*, **19**, 445–71.

Felsenstein, J. (1988*b*). Phylogenies from molecular sequences: Inference and reliability. *Annual Review of Genetics*, **22**, 521–65.

Felsenstein, J. (1988*c*). Perils of molecular introspection. *Nature*, **335**, 118.

Felsenstein, J. (1989). *PHYLIP 3.2 manual*. University of California Herbarium, Berkeley, California.

Felsenstein, J. (1991). Counting phylogenetic invariants in some simple cases. *Journal of Theoretical Biology*, **152**, 357–76.

Feng, D. F. and Doolittle, R. F. (1987). Progressive sequence alignment as a prerequisite to correct phylogenetic trees. *Journal of Molecular Evolution*, **25**, 351–60.

Fink, W. L. (1982). The conceptual relationship between ontogeny and phylogeny. *Paleobiology*, **8**, 254–64.

Fitch, W. M. (1966). An improved method of testing for evolutionary homology. *Journal of Molecular Biology*, **16**, 9–16.

Fitch, W. M. (1970). Distinguishing homologous from analogous proteins. *Systematic Zoology*, **19**, 99–113.

Fitch, W. M. (1971). Toward defining the course of evolution: minimum change for a specified tree topology. *Systematic Zoology*, **20**, 406–16.

Fitch, W. M. (1977). The phyletic interpretation of macromolecular sequence information: simple methods. In *Major patterns in vertebrate evolution* (ed. M. K. Hecht, P. C. Goody, and B. N. Hecht), pp. 169–204. Plenum Press, New York.

Fitch, W. M. (1981). A non-sequential method for constructing trees and hierarchical classifications. *Journal of Molecular Evolution*, **18**, 30–7.

Fitch, W. M. (1984). Cladistics and other methods: problems, pitfalls and potentials. In *Cladistics: perspectives on the reconstruction of evolutionary history* (ed. T. Duncan and T. F. Stuessy), pp. 221–52. Columbia University Press, New York.

Fitch, W. M. and Margoliash, E. (1967). Construction of phylogenetic trees. *Science*, **155**, 279–84.

Fitch, W. M. and Markowitz, E. (1970). An improved method for determining codon variability in a gene and its application to the rate of fixations of mutations in evolution. *Biochemical Genetics*, **4**, 579–93.

Fitch, W. M. and Yasunobu, K. T. (1975). Phylogenies from amino acid sequences with gaps. *Journal of Molecular Evolution*, **5**, 1–24.

Fortey, R. A. and Jefferies, R. P. S. (1982). Fossils and phylogeny – a compromise approach. In *Problems in phylogenetic reconstruction* (ed. K. A. Joysey and A. E. Friday), pp. 197–234. Academic Press, London.

Gauld, I. and Underwood, G. (1986). Some applications of the Le Quesne compatibility test. *Biological Journal of the Linnean Society*, **29**, 191–222.

Gauthier, J., Kluge, A. G., and Rowe, T. (1988). Amniote phylogeny and the importance of fossils. *Cladistics*, **4**, 105–209.

Glazer, A. N. (1987). Phycobilisomes: assembly and attachment. In *The cyanobacteria* (ed. P. Fay and C. Van Baalen), pp. 69–94. Elsevier, Amsterdam.

Goloboff, P. A. (1991). Homoplasy and the choice among cladograms. *Cladistics*, **7**, 215–32.

Goodman, M., Czelusniak, J., Moore, G. W., Romero-Herrera, A. E., and Matsuda, G. (1979). Fitting the gene lineage into its species lineage. *Systematic Zoology*, **28**, 132–67.

Goodman, M., Miyamoto, M. M., and Czelusniak, J. (1987). Pattern and process in vertebrate phylogeny revealed by coevolution of molecules and morphologies. In *Molecules and morphology in evolution: conflict or compromise?* (ed. C. Patterson), pp. 141–76. Cambridge University Press, Cambridge.

Graham, R. L. and Foulds, L. R. (1982). Unlikelihood that minimal phylogenies for a realistic biological study can be constructed in reasonable computational time. *Mathematical Biosciences*, **60**, 133–42.

Gray, G. S. and Fitch, W. M. (1983). Evolution of antibiotic resistance genes: the DNA sequence of a kanamycin resistance gene from *Staphylococcus aureus*. *Molecular Biology and Evolution*, **1**, 57–66.

Griffiths, G. C. D. (1974). On the foundation of biological systematics. *Acta Biotheoretica*, **23**, 85–131.

Haeckel, E. (1866). *Generelle Morphologie der Organismen*. G. Reimer, Berlin.

Hammond, P. M. (1979). Wing-fold mechanisms of beetles, with special reference to investigations of adephagan phylogeny (Coleoptera). In *Carabid beetles: their evolution, natural history and classification*, Proceedings of the First International Symposium of Carabidology, (ed. T. L. Erwin, G. E. Ball, and D. R. Whitehead), pp. 113–80. W. Junk, The Hague.

Hartigan, J. A. (1972). Minimum mutation fits to a given tree. *Biometrics*, **29**, 53–65.

Hay, W. W. (1972). Probabilistic stratigraphy. *Ecologae Geologae Helvetica*, **65**, 255–66.

Hein, J. (1989*a*). A new method that simultaneously aligns and reconstructs ancestral sequences for any number of homologous sequences when a phylogeny is given. *Molecular Biology and Evolution*, **6**, 649–68.

Hein, J. (1989*b*). A tree reconstruction method that is economical in the number of pairwise comparisons used. *Molecular Biology and Evolution*, **6**, 669–84.

Henderson, I. (1989). Quantitative biogeography: an investigation into concepts and methods. *New Zealand Journal of Zoology*, **16**, 495–510.

Hendy, M. D. and Penny, D. (1982). Branch and bound algorithms to determine minimal evolutionary trees. *Mathematical Biosciences*, **59**, 277–90.

Hendy, M. D. and Penny, D. (1989). A framework for the quantitative study of evolutionary trees. *Systematic Zoology*, **38**, 297–309.

Hennig, W. (1950). *Grundzüge einer Theorie der phylogenetischen Systematik*. Deutsche Zentralverlag, Berlin.

Hennig, W. (1966). *Phylogenetic systematics*. University of Illinois Press, Urbana, Illinois.

Hennig, W. (1969). *Die Stammesgeschichte der Insekten*. E. Kramer, Frankfurt am Main.

Hennig, W. (1983). Stammesgeschichte der Chordaten. *Forschritte in der zoologischen Systematik und Evolutionsforschung*, **2**, 1–208.

Higgins, D. G. and Sharp, P. M. (1988). CLUSTAL: a package for performing multiple sequence alignment on a microcomputer. *Gene*, **73**, 237–44.

Higgins, D. G. and Sharp, P. M. (1989). Fast and sensitive multiple sequence alignments on a microcomputer. *CABIOS*, **5**, 151–3.

Hill, C. R. and Camus, J. M. (1986). Evolutionary cladistics of marrattialean ferns. *Bulletin of the British Museum (Natural History)*, (Botany), **14**, 219–300.

Hillis, D. M. (1987). Molecular versus morphological approaches to systematics. *Annual Review of Ecology and Systematics*, **18**, 23–42.

Hillis, D. M. and Moritz, C. (1990). An overview of applications of molecular systematics. In *Molecular systematics* (ed. D. M. Hillis and C. Moritz), pp. 502–15. Sinauer Associates, Sunderland, Massachusetts.

Holmquist, R. (1972). Theoretical foundations for quantitative paleogenetics. Part 1: DNA. *Journal of Molecular Evolution*, **1**, 115–33.

Hull, D. L. (1988). *Science as a process: an evolutionary account of the social and conceptual development of science*. University of Chicago Press, Chicago.

Humphries, C. J. and Funk, V. A. (1984). Cladistic methodology. In *Current concepts in plant taxonomy* (ed. V. H. Heywood and D. M. Moore), pp. 323–62. Academic Press, London.

Humphries, C. J., Ladiges, P. Y., Roos, M., and Zandee, M. (1988). Cladistic biogeography. In *Analytical biogeography: an integrated approach to the study of animal and plant distributions* (ed. A. Myers and P. Giller), pp. 371–404. Chapman and Hall, London.

Humphries, C. J. and Parenti, L. R. (1986). *Cladistic biogeography*. The Clarendon Press, Oxford.

Janvier, P. (1984). Cladistics: theory, purpose and evolutionary implications. In *Evolutionary theory paths into the future* (ed. J. W. Pollard), pp. 39–75. Wiley Interscience, New York.

Jefferies, R. P. S. (1979). The origin of the chordates – a methodological essay. In *The origin of the major vertebrate groups* (ed. M. R. House), pp. 443–77. Academic Press, London.

Jones, K. (1974). Chromosome evolution by Robertsonian change in *Gibasis* (Commelinaceae). *Chromosoma*, **45**, 353–68.

Jones, K. (1977). The model of Robertsonian change in karyotype evolution in higher plants. *Chromosomes Today*, **6**, 121–9.

Jong, R. de (1980). Some tools for evolutionary and phylogenetic studies. *Zeitschrift für zoologische Systematik und Evolutionsforschung*, **18**, 1–23.

Jukes, T. H. and Cantor, C. R. (1969). Evolution of protein molecules. In *Mammalian protein metabolism* (ed. H. Munro), pp. 21–132. Academic Press, New York.

Kimura, M. (1969). The rate of evolution considered from the standpoint of population genetics. *Proceedings of the National Academy of Science USA*, **63**, 1181–8.

Kimura, M. (1980). A simple method for estimating evolutionary rate of base substitutions through comparative studies of nucleotide sequences. *Journal of Molecular Evolution*, **16**, 111–20.

Kimura, M. (1983). *The neutral theory of molecular evolution*. Cambridge University Press, Cambridge.

Klotz, L. C. and Blanken, R. L. (1981). A practical method for calculating evolutionary trees from sequence data. *Journal of Theoretical Biology*, **91**, 261–72.

Klotz, L. C., Komar, N., Blanken, R. L., and Mitchell, R. M. (1979). Calculation of evolutionary trees from sequence data. *Proceedings of the National Academy of Science USA*, **76**, 4516–20.

Kluge, A. G. (1985). Ontogeny and phylogenetic systematics. *Cladistics*, **1**, 13–27.

Kluge, A. G. (1988*a*). The characteristics of ontogeny. In *Ontogeny and systematics* (ed. C. J. Humphries), pp. 57–82. Columbia University Press, New York.

Kluge, A. G. (1988*b*). Parsimony in vicariance biogeography: a quantitative method and a Greater Antillean example. *Systematic Zoology*, **37**, 315–28.

Kluge, A. G. and Farris, J. S. (1969). Quantitative phyletics and the evolution of anurans. *Systematic Zoology*, **18**, 1–32.

Kluge, A. G. and Strauss, R. E. (1985). Ontogeny and systematics. *Annual Review of Ecology and Systematics*, **16**, 247–68.

Kociolek, J. P. and Williams, D. M. (1987). Unicell ontogeny and phylogeny: examples from the diatoms. *Cladistics*, **3**, 274–84.

Konings, D. A., Hogweg, M. P., and Hesper, B. (1987). Evolution of the primary and secondary structures of the E1a mRNAs of the adenovirus. *Molecular Biology and Evolution*, **4**, 300–14.

Kraus, F. (1988). An empirical evaluation of the use of the ontogeny polarisation criterion in phylogenetic inference. *Systematic Zoology*, **37**, 106–41.

Lake, J. A. (1987a). A rate independent technique for analysis of nucleic acid sequences: Evolutionary parsimony. *Molecular Biology and Evolution*, **4**, 167–91.

Lake, J. A. (1987b). Determining evolutionary distances from highly derived nucleic acid sequences: operator metrics. *Journal of Molecular Evolution*, **26**, 59–73.

Lauder, G. V. (1986). Homology, analogy, and the evolution of behavior. In *Evolution of animal behavior* (ed. M. H. Nitecki and J. A. Kitchell), pp. 9–40. Oxford University Press, Oxford.

Lauder, G. V. (1990). Functional morphology and systematics: studying functional patterns in an historical context. *Annual Review of Ecology and Systematics*, **21**, 317–40.

Le Quesne, W. J. (1969). A method of selection of characters in numerical taxonomy. *Systematic Zoology*, **18**, 201–5.

Li, W.-H. (1981). A simple method for constructing phylogenetic trees from distance matrices. *Proceedings of the National Academy of Sciences USA*, **78**, 1085–9.

Li, W.-H. and Graur, D. (1991). *Fundamentals of molecular evolution*. Sinauer Associates, Sunderland, Massachusetts.

Li, W.-H. and Nei, M. (1990). Limitations of the evolutionary parsimony method of phylogenetic analysis. *Molecular Biology and Evolution*, **6**, 82–102.

Li, W.-H. and Tanimura, M. (1987). The molecular clock runs more slowly in man than in apes and monkeys. *Nature*, **326**, 93–6.

Li, W.-H., Wolfe, K. H., Sourdis, J., and Sharp, P. M. (1987). Reconstruction of phylogenetic trees and estimation of divergence times under nonconstant rates of evolution. *Cold Spring Harbor Symposia on Quantitative Biology*, **52**, 847–56.

Lipscomb, D. (1985). The eukaryotic kingdoms. *Cladistics*, **1**, 127–40.

Lipscomb, D. L. (1990). Two methods for calculating cladogram characters: transformation series analysis and the iterative FIG/FOG method. *Systematic Zoology*, **39**, 277–88.

Løvtrup, S. (1977). *The phylogeny of the vertebrata*. Wiley, London.

Løvtrup, S. (1978). On Von Baerian and Haeckelian recapitulation. *Systematic Zoology*, **27**, 348–52.

Lundberg, J. G. (1972). Wagner networks and ancestors. *Systematic Zoology*, **21**, 398–413.

Lundberg, J. G. (1973). More on primitiveness, higher level phylogenies and ontogenetic transformations. *Systematic Zoology*, **22**, 327–9.

Mabee, P. M. (1989). An empirical rejection of the ontogenetic polarity criterion. *Cladistics*, **5**, 409–16.

Maddison, W. P. (1989). Reconstructing character evolution on polytomous cladograms. *Cladistics*, **5**, 365–77.

Maddison, W. P., Donoghue, M. J., and Maddison, D. R. (1984). Outgroup analysis and parsimony. *Systematic Zoology*, **33**, 83–103.

Mayr, E. (1969). *Principles of systematic zoology*. McGraw Hill, New York.

McKenna, M. C. (1975). Towards a phylogenetic classification of the Mammalia. In *Phylogeny of the primates* (ed. W. P. Luckett and F. S. Szalay), pp. 21–46. Plenum Press, New York.

Meacham, C. A. (1984). The role of hypothesized direction of characters in the estimation of evolutionary history. *Taxon*, **33**, 26–38.

Meacham, C. A. and Estabrook, G. F. (1985). Compatibility methods in systematics. *Annual Review of Ecology and Systematics*, **16**, 431–46.

Mickevich, M. F. (1982). Transformation series analysis. *Systematic Zoology*, **31**, 461–78.

Mickevich, M. F. and Lipscomb, D. (1991). Parsimony and the choice between different transformations for the same character set. *Cladistics*, **7**, 111–39.

Mickevich, M. F. and Weller, S. J. (1990). Evolutionary character analysis: tracing character change on a cladogram. *Cladistics*, **6**, 137–70.

Miyamoto, M. M. (1985). Consensus cladograms and general classifications. *Cladistics*, **1**, 186–9.

Miyamoto, M. M. and Boyle, S. M. (1989). The potential importance of mitochondrial DNA sequence data to eutherian mammal phylogeny. In *The hierarchy of life* (ed. B. Fernholm, K. Bremer, and H. Jörnvall), pp. 437–50. Elsevier, Amsterdam.

Miyazaki, J. M. and Mickevich, M. F. (1982). Evolution of *Chesapecten* (Mollusca: Bivalvia, Miocene-Pliocene) and the biogenetic law. *Evolutionary Biology*, **15**, 369–410.

Mooi, R. (1989). The outgroup criterion revisited via naked zones and alleles. *Systematic Zoology*, **38**, 283–90.

Moritz, C. and Hillis, D. M. (1990). Molecular systematics: context and controversies. In *Molecular Systematics* (ed. D. M. Hillis, and C. Moritz), pp. 1–10. Sinauer Associates, Sunderland, Massachusetts.

Muir-Wood, H. M. (1955). *A history of the classification of the phylum Brachiopoda*. British Museum (Natural History), London.

Nanney, D. L., Preparata, R. M., Preparata, F. P., Meyer, E. B., and Simon, E. M. (1989). Shifting ditypic site analysis: heuristics for expanding the phylogenetic range of nucleotide sequences in Sankoff analyses. *Journal of Molecular Evolution*, **28**, 451–9.

Needleman, S. B. and Wunsch, C. D. (1970). A general method applicable to the search for similarities in the amino acid sequence of two proteins. *Journal of Molecular Biology*, **48**, 443–53.

Neff, N. A. (1986). A rational basis for a priori character weighting. *Systematic Zoology*, **35**, 110–23.

Nei, M. (1987). *Molecular evolutionary genetics*. Columbia University Press, New York.

Nei, M., Tajima, F., and Tateno, Y. (1983). Accuracy of phylogenetic trees from molecular data. II. Gene frequency data. *Journal of Molecular Evolution*, **19**, 153–70.

Nelson, G. J. (1972). Phylogenetic relationships and classification. *Systematic Zoology*, **21**, 227–30.

Nelson, G. J. (1973a). The higher-level phylogeny of vertebrates. *Systematic Zoology*, **22**, 87–91.

Nelson, G. J. (1973b). Negative gains and positive losses: a reply to J. C. Lundberg. *Systematic Zoology*, **22**, 330.

Nelson, G. J. (1973*c*). Classification as an expression of phylogenetic relationships. *Systematic Zoology*, **22**, 344–59.

Nelson, G. J. (1978). Ontogeny, phylogeny, paleontology, and the biogenetic law. *Systematic Zoology*, **27**, 324–45.

Nelson, G. J. (1979). Cladistic analysis and synthesis: Principles and definitions with a historical note on Adanson's *Familles des Plantes* (1763–1764). *Systematic Zoology*, **28**, 1–21.

Nelson, G. J. (1982). Cladistique et biogeographie. *Comptes Rendue de la Société de Biogéographie*, **58**, 75–94.

Nelson, G. J. (1984). Cladistics and biogeography. In *Cladistics: perspectives on the reconstruction of evolutionary history* (eds T. Duncan and T. F. Stuessy), pp. 273–93. Columbia University Press, New York.

Nelson, G. J. (1985). Outgroups and ontogeny. *Cladistics*, **1**, 29–45.

Nelson, G. J. (1989). Cladistics and evolutionary models. *Cladistics*, **5**, 275–89.

Nelson, G. J. and Platnick, N. I. (1981). *Systematics and biogeography: cladistics and vicariance*. Columbia University Press, New York.

Nelson, G. and Platnick, N. I. (1991). Three-taxon statements: a more precise use of parsimony? *Cladistics*, **7**, 351–66.

Newell, N. D. (1959). The nature of the fossil record. *Proceedings of the American Philosophical Society*, **103**, 264–85.

Nixon, K. C. and Wheeler, Q. D. The concept of character adjacency. *Cladistics* (In preparation).

O'Grady, R. T. (1985). Ontogenetic sequences and the phylogenetics of parasitic flatworm life cycles. *Cladistics*, **1**, 159–70.

Olsen, G. J. (1988). Phylogenetic analysis using ribosomal RNA. *Methods in Enzymology*, **164**, 793–812.

Owen, R. (1843). *Lectures on the comparative anatomy and physiology of the invertebrate animals*. Longman, Brown, Green, and Longmans, London.

Page, R. D. M. (1987). Graphs and generalized tracks: quantifying Croizat's panbiogeography. *Systematic Zoology*, **36**, 1–12.

Page, R. D. M. (1988). Quantitative cladistic biogeography: Constructing and comparing area cladograms. *Systematic Zoology*, **37**, 254–70.

Page, R. D. M. (1989*a*). Comments on component-compatibility in historical biogeography. *Cladistics*, **5**, 167–82.

Page, R. D. M. (1989*b*). *COMPONENT. Users manual and software. Release 1.5.* Published by the author.

Page, R. D. M. (1990). Temporal congruence and cladistic analysis of biogeography and cospeciation. *Systematic Zoology*, **39**, 205–26.

Page, R. D. M. (1991). Clocks, clades, and cospeciation: comparing rates of evolution and timing of cospeciation events in host – parasite assemblages. *Systematic Zoology*, **40**, 188–98.

Patterson, C. (1977). The contribution of paleontology to teleostean phylogeny. In *Major patterns in vertebrate evolution* (ed. M. K. Hecht, P. C. Goody, and B. M. Hecht), pp. 579–643. Plenum Press, New York.

Patterson, C. (1978). Verifiability in systematics. *Systematic Zoology*, **27**, 218–22.

Patterson, C. (1980). Cladistics. *The Biologist*, **27**, 234–9.

Patterson, C. (1981*a*). Significance of fossils in determining evolutionary relationships. *Annual Review of Ecology and Systematics*, **12**, 195–223.

Patterson, C. (1981*b*). Methods in paleobiogeography. In *Vicariance biogeography: a critique* (ed. G. Nelson and D. E. Rosen), pp. 446–89. Columbia University Press, New York.

Patterson, C. (1982*a*). Morphological characters and homology. In *Problems in phylogenetic reconstruction* (ed. K. A. Joysey and A. E. Friday), pp. 21–74. Academic Press, London.

Patterson, C. (1982*b*). Cladistics and classification. *New Scientist*, **94**, 303–6.

Patterson, C. (1983). How does phylogeny differ from ontogeny? In *Development and evolution* (ed. B. C. Goodwin, H. Holder, and C. C. Wylie), pp. 1–31. Cambridge University Press, Cambridge.

Patterson, C. (1988*a*). Homology in classical and molecular biology. *Molecular Biology and Evolution*, **5**, 603–25.

Patterson, C. (1988*b*). The impact of evolutionary theories on systematics. In *Prospects in systematics* (ed. D. L. Hawksworth), pp. 59–91. Clarendon Press, Oxford.

Patterson, C. (1989). Phylogenetic relations of major groups: Conclusions and prospects. In *The hierarchy of life* (ed. B. Fernholm, K. Bremer, and H. Jörnvall), pp. 471–88. Elsevier, Amsterdam.

Patterson, C. and Rosen, D. E. (1977). Review of ichthyodectiform and other Mesozoic teleost fishes and the theory and practice of classifying fossils. *Bulletin of the American Museum of Natural History*, **158**, 81–172.

Paul, C. R. C. (1982). The adequacy of the fossil record. In *Problems in phylogenetic reconstruction* (ed. K. A. Joysey and A. E. Friday), pp. 75–117. Academic Press, London.

Penny, D. (1988). What was the first living cell? *Nature*, **331**, 111–12.

Pimentel, R. A. and Riggins, R. (1987). The nature of cladistic data. *Cladistics*, **3**, 201–9.

Platnick, N. I. (1979). Philosophy and the transformation of cladistics. *Systematic Zoology*, **28**, 537–46.

Platnick, N. I. (1981). Widespread taxa and biogeographic congruence. *Advances in Cladistics*, **1**, 223–7.

Platnick, N. I. and Cameron, H. D. (1977). Cladistic methods in textual, linguistic, and phylogenetic analysis. *Systematic Zoology*, **26**, 380–5.

Platnick, N. I. and Nelson, G. J. (1978). A method of analysis for historical biogeography. *Systematic Zoology*, **27**, 1–16.

Platnick, N. I., Griswold, C. E., and Coddington, J. A. (1991). On missing entries in cladistic analysis. *Cladistics*, **7**, 337–43.

Riedl, R. (1979). *Order in living organisms*. John Wiley, New York.

Rieppel, O. (1985). Ontogeny and the hierarchy of types. *Cladistics*, **1**, 234–46.

Rieppel, O. (1988). *Fundamentals of comparative biology*. Birkhauser Verlag, Berlin.

Rogers, J. S. (1986). Deriving phylogenetic trees from allele frequencies: a comparison of nine genetic distances. *Systematic Zoology*, **35**, 297–310.

Rolf, J. (1988). *NTSYS-PC: numerical taxonomy and multivariate analysis system*, version 1.4. Exeter, New York.

Rosen, D. E. (1976). A vicariance model of Caribbean biogeography. *Systematic Zoology*, **24**, 431–64.

Rosen, D. E. (1978). Vicariant patterns and historical explanations in biogeography. *Systematic Zoology*, **27**, 159–88.

Rosen, D. E. (1979). Fishes from the uplands and intermontane basins of Guatemala: revisionary studies and comparative geography. *Bulletin of the American Museum of Natural History*, **162**, 267–376.

Rosen, D. E. (1982). Do current theories of evolution satisfy the basic requirements of explanation? *Systematic Zoology*, **31**, 76–85.

Rosen, D. E. (1984). Hierarchies and history. In *Evolutionary theory: paths into the future* (ed. J. W. Pollard), pp. 77–97. John Wiley, New York.

Rosen, D. E., Forey, P. L., Gardiner, B. G., and Patterson, C. (1981). Lungfishes, tetrapods, paleontology and plesiomorphy. *Bulletin of the American Museum of Natural History*, **167**, 159–276.

Ross, H. H. (1950). *A textbook of entomology*. (2nd edn). John Wiley, New York.

Ross, H. H. (1974). *Biological systematics*. Addison-Wesley, Reading, Massachusetts.

Saether, O. A. (1979). Underlying synapomorphies and anagenetic analysis. *Zoologica Scripta*, **8**, 305–12.

Saether, O. A. (1983). The canalized evolutionary potential: Inconsistencies in phylogenetic reasoning. *Systematic Zoology*, **32**, 343–59.

Saether, O. A. (1986). The myth of objectivity – post-Hennigian deviations. *Cladistics*, **2**, 1–13.

Saitou, N. and Nei, M. (1987). The neighbor-joining method: a new method for reconstructing phylogenetic trees. *Molecular Biology and Evolution*, **4**, 406–25.

Sanderson, M. J. (1989). Confidence limits on phylogenies: the bootstrap revisited. *Cladistics*, **5**, 113–29.

Sanderson, M. J. and Donoghue, M. J. (1989). Patterns of variation in levels of homoplasy. *Evolution*, **43**, 1781–95.

Sankoff, D. (1975). Minimal mutation trees of sequences. *SIAM Journal of Applied Mathematics*, **21**, 35–42.

Sankoff, D. and Cedergren, R. J. (1983). Simultaneous comparison of three or more sequences related by a tree. In *Time warps, string edits and macromolecules: the theory and practice of sequence comparison* (ed. D. Sankoff and J. B. Kruskall), pp. 253–63. Addison-Wesley, Reading, Massachusetts.

Sarich, V. M. (1969). Pinniped origins and the rate of evolution of carnivore albumins. *Systematic Zoology*, **18**, 286–95.

Sattath, S. and Tversky, A. (1977). Additive similarity trees. *Psychometrika*, **42**, 319–45.

Scherer, S. (1990). The protein molecular clock. Time for a re-evaluation. *Evolutionary Biology*, **24**, 83–106.

Schoch, R. M. (1986). *Phylogeny reconstruction in paleontology*. Van Nostrand Reinhold, New York.

Schultze, H.-P. (1987). Dipnoans as sarcopterygians. In *The biology and evolution of lungfishes* (ed. W. E. Bemis, W. W. Burggren, and N. E. Kemp), pp. 39–74. Alan Liss, New York.

Sellers, P. (1974). On the theory and computation of evolutionary distances. *SIAM Journal of Applied Mathematics*, **26**, 787–93.

Sharkey, M. J. (1989). A hypothesis-independent method of character weighting for cladistic analysis. *Cladistics*, **5**, 63–86.

Sidow, A. and Wilson, A. C. (1990). Compositional statistics: an improvement of evolutionary parsimony and its application to deep branches in the tree of life. *Journal of Molecular Evolution*, **31**, 51–68.

Simpson, G. G. (1945). The principles of classification and a classification of mammals. *Bulletin of the American Museum of Natural History*, **85**, 1–350.

Simpson, G. G. (1961). *Principles of animal taxonomy*. Columbia University Press, New York.

Smith, A. B. (1984). *Echinoid palaeobiology*. George Allen and Unwin, London.

Smith, A. B. (1989). RNA sequence data in phylogenetic reconstruction: testing the limits of its resolution. *Cladistics*, **5**, 321–44.

Smith, T. F. and Fitch, W. M. (1981). Comparative biosequence metrics. *Journal of Molecular Evolution*, **18**, 36–46.

Sneath, P. H. A. and Sokal, R. R. (1973). *Numerical taxonomy. The principles and practice of numerical classification*. W. H. Freeman, San Francisco.

Sokal, R. R. and Sneath, P. H. A. (1963). *Principles of numerical taxonomy*. W. H. Freeman, San Francisco.

Springer, M. and Krajewski, C. (1989). DNA hybridisation in animal taxonomy: a critique from first principles. *Quarterly Review of Biology*, **64**, 291–318.

Stacey, S. N., Lansman, R. A., Brock, H. W., and Grigliatti, T. A. (1986). Distribution and conservation of mobile elements in the genus *Drosophila*. *Molecular Biology and Evolution*, **3**, 522–34.

Steele, K. P., Holsinger, K. E., Jansen, R. K., and Taylor, D. W. (1991). Assessing the reliability of 5S rRNA sequence data for phylogenetic analysis in green plants. *Molecular Biology and Evolution*, **8**, 240–8.

Stevens, P. F. (1980). Evolutionary polarity of character states. *Annual Review of Ecology and Systematics*, **11**, 333–58.

Stevens, P. F. (1991). Character states, continuous variation and phylogenetic analysis: a review. *Systematic Botany*, **16**, 553–83.

Stryer, L. (1981). *Biochemistry*. (2nd edn). Freeman, San Francisco.

Studier, J. A. and Keppler, K. J. (1988). A note on the neighbor-joining algorithm of Saitou and Nei. *Molecular Biology and Evolution*, **5**, 729–31.

Swofford, D. L. (1981). On the utility of the distance Wagner procedure. *Advances in Cladistics*, **1**, 25–44.

Swofford, D. L. (1990). *PAUP: Phylogenetic analysis using parsimony version 3.0*. Illinois Natural History Survey, Champaign, Illinois.

Swofford, D. L. and Berlocher, S. H. (1987). Inferring evolutionary trees from gene frequency data under the principle of maximum parsimony. *Systematic Zoology*, **36**, 293–325.

Swofford, D. L. and Maddison, W. P. (1987). Reconstructing ancestral character states under Wagner parsimony. *Mathematical Biosciences*, **87**, 199–229.

Swofford, D. L. and Olsen, G. J. (1990). Phylogeny reconstruction. In *Molecular Systematics* (ed. D. M. Hillis and C. Moritz), pp. 411–501. Sinauer, Sunderland, Massachusetts.

Tajima, F. and Nei, M. (1984). Estimation of evolutionary distance between nucleotide sequences. *Molecular Biology and Evolution*, **1**, 269–85.

Tateno, Y. (1990). A method for molecular phylogeny construction by direct use of nucleotide sequence data. *Journal of Molecular Evolution*, **30**, 85–93.

Tateno, Y., Nei, M., and Tajima, F. (1981). Accuracy of estimated phylogenetic trees from molecular data. I. Distantly related species. *Journal of Molecular Evolution*, **18**, 387–404.

Thorpe, J. P. (1982). The molecular clock hypothesis: biochemical evolution, genetic differentiation and systematics. *Annual Review of Ecology and Systematics*, **13**, 139–68.

Trueb, L. and Cloutier, R. (1991). A phylogenetic investigation of the inter- and intra-relationships of the Lissamphibian (Amphibia: Temnospondyli). In *Origins of the major groups of tetrapods: controversies and consensus* (ed. H.-P. Schultze and L. Trueb), pp. 223–313. Cornell University Press, Ithaca.

Turner, J. R. G. (1983). Mimetic butterflies and punctuated equilibria: some old light on a new paradigm. *Biological Journal of the Linnean Society*, **20**, 277–300.

Turner, J. R. G. (1984). Mimicry: the palatability spectrum and its consequences. In *The biology of butterflies*, Symposium of the Royal Entomological Society of London (ed. R. I. Vane-Wright and P. R. Ackery), pp. 141–61. Academic Press, London.

Tyler, S. (1988). The role of function in determination of homology and convergence – examples from invertebrate adhesive organs. *Forschritte der Zoologie*, **36**, 331–47.

Uy, R. and Wold, F. (1977). Posttranslational covalent modification of proteins. *Science*, **198**, 890–96.

Vane-Wright, R. I. (1979). Towards a theory of evolution of butterfly colour patterns under directional and disruptive selection. *Biological Journal of the Linnean Society*, **11**, 141–52.

Wagner, W. H. Jr. (1961). Problems in the classification of ferns. *Recent Advances in Botany*, **1**, 841–4.

Wagner, W. H. Jr. (1963). Biosystematics and taxonomic categories in lower vascular plants. *Regnum Vegetabile*, **27**, 63–71.

Waterman, M. S., Smith, T. F., Singh, M., and Beyer, W. A. (1977). Additive evolutionary trees. *Journal of Theoretical Biology*, **64**, 199–213.

Watrous, L. E. and Wheeler, Q. D. (1981). The outgroup comparison method of character analysis. *Systematic Zoology*, **30**, 1–11.

West, J. G. and Faith, D. P. (1990). Data, methods and assumptions in phylogenetic inference. *Australian Systematic Botany*, **3**, 9–20.

Weston, P. H. (1988). Indirect and direct methods in systematics. In *Ontogeny and systematics* (ed. C. J. Humphries), pp. 27–56. Columbia University Press, New York.

Wheeler, Q. D. (1986). Character weighting and cladistic analysis. *Systematic Zoology*, **35**, 102–9.

Wheeler, Q. D. (1990). Ontogeny and character phylogeny. *Cladistics*, **6**, 225–68.

Wheeler, W. C. (1990). Nucleic acid sequence phylogeny and random outgroups. *Cladistics*, **6**, 363–7.

Wheeler, W. C. and Honeycutt, R. L. (1988). Paired sequence differences in ribosomal RNAs: evolutionary and phylogenetic implications. *Molecular Biology and Evolution*, **5**, 90–6.

Wilbur, W. J. and Lipman, D. J. (1983). Rapid similarity searches of nucleic acid and protein data banks. *Proceedings of the National Academy of Science USA*, **80**, 726–30.

Wiley, E. O. (1979). The annotated Linnean hierarchy, with comments on natural taxa and competing systems. *Systematic Zoology*, **28**, 308–37.

Wiley, E. O. (1980). Phylogenetic systematics and vicariance biogeography. *Systematic Botany*, **5**, 194–220.

Wiley, E. O. (1981). *Phylogenetics: the theory and practice of phylogenetic systematics*. Wiley Interscience, New York.

Wiley, E. O. (1987). Methods in vicariance biogeography. In *Systematics and evolution: a matter of diversity* (ed. P. Hovencamp, E. Gittenberg, E. Hennipman, R. de Jong, M. C. Roos, R. Sluys, and M. Zandee), pp. 283–306. IPTS Faculty of Biology, Utrecht University, The Netherlands.

Wiley, E. O. (1988). Parsimony analysis and vicariance biogeography. *Systematic Zoology*, **37**, 271–90.

Wiley, E. O., Siegel-Causey, D. J., Brooks, D. R., and Funk, V. A. (1991). *The compleat cladist: a primer of phylogenetic procedures*. Museum of Natural History, University of Kansas, Lawrence.

Williams, D. M., Scotland, R. W., and Blackmore, S. (1990). Is there a direct ontogenetic criterion in systematics? *Biological Journal of the Linnean Society*, **39**, 99–108.

Williams, P. L. and Fitch, W. M. (1989). Finding the minimal change in a given tree. In *The hierarchy of life* (ed. B. Fernholm, K. Bremer, and H. Jörnvall), pp. 453–70. Elsevier, Amsterdam.

Williams, P. L. and Fitch, W. M. (1990). Phylogeny determination using dynamically weighted parsimony methods. *Methods in Enzymology*, **183**, 615–26.

Zandee, M. and Roos, M. C. (1987). Component-compatibility in historical biogeography. *Cladistics*, **3**, 305–32.

Zuckerkandl, E. and Pauling, L. (1965). Molecules as documents of evolutionary history. *Journal of Theoretical Biology*, **8**, 357–66.

Systematics Association
Special Volumes

1. The new systematics (1940)
   Edited by J. S. Huxley (Reprinted (1971)
2. Chemotaxonomy and serotaxonomy (1968)*
   Edited by J. G. Hawkes
3. Data processing in biology and geology (1971)*
   Edited by J. L. Cutbill
4. Scanning electron microscopy (1971)*
   Edited by V. H. Heywood
   Out of print
5. Taxonomy and ecology (1973)*
   Edited by V. H. Heywood
6. The changing flora and fauna of Britain (1974)*
   Edited by D. L. Hawksworth
   Out of print
7. Biological identification with computers (1975)*
   Edited by R. J. Pankhurst
8. Lichenology: progress and problems (1976)*
   Edited by D. H. Brown, D. L. Hawksworth, and R. H. Bailey
9. Key works to the fauna and flora of the British Isles and north-western
   Europe, *4th edition* (1978)*
   Edited by G. J. Kerrich, D. L. Hawksworth, and R. W. Sims
10. Modern approaches to the taxonomy of red and brown algae (1978)*
    Edited by D. E. G. Irvine and J. H. Price
11. Biology and systematics of colonial organisms (1979)*
    Edited by G. Larwood and B. R. Rosen
12. The origin of major invertebrate groups (1979)*
    Edited by M. R. House
13. Advances in bryozoology (1979)*
    Edited by G. P. Larwood and M. B. Abbot
14. Bryophyte systematics (1979)*
    Edited by G. C. S. Clarke and J. G. Duckett
15. The terrestrial environment and the origin of land vertebrates (1980)*
    Edited by A. L. Panchen

37. Biosystematics of haematophagous insects (1988)‡
    Edited by M. W. Service
38. The chromophyte algae: problems and perspectives (1989)‡
    Edited by J. C. Green, B. S. C. Leadbeater, and W. Diver
39. Electrophoretic studies on agricultural pests (1989)‡
    Edited by Hugh D. Loxdale and J. den Hollander
40. Evolution, systematics, and fossil history of the Hamamelidae (2 Volumes) (1989)‡
    Edited by Peter R. Crane and Stephen Blackmore
41. Scanning electron microscopy in taxonomy and functional morphology (1990)‡
    Edited by D. Claugher
42. Major evolutionary radiations (1990)‡
    Edited by P. D. Taylor and G. P. Larwood
43. Tropical lichens: their systematics, conservation, and ecology (1991)‡
    Edited by D. J. Galloway
44. Pollen and spores: patterns of diversification‡
    Edited by S. Blackmore and S. H. Barnes
45. The biology of free-living heterotrophic flagellates‡
    Edited by D. J. Patterson and J. Larsen
46. Plant-animal interactions in the marine benthos‡
    Edited by D. M. John, S. J. Hawkins, and J. H. Price

* Published by Academic Press for the Systematics Association
† Published by the Palaeontological Association in conjunction with the Systematics Association
‡Published by the Oxford University Press for the Systematics Association

Index